Designing usable electronic text

Dedication
To err is human, to forgive design

This book is dedicated to my professional colleagues, past and present at the HUSAT Research Institute, Loughborough University of Technology, England.

Designing usable electronic text
Ergonomic aspects of human information usage

Andrew Dillon
Indiana University
USA

CRC Press
Taylor & Francis Group
Boca Raton London New York

CRC Press is an imprint of the
Taylor & Francis Group, an **informa** business
A TAYLOR & FRANCIS BOOK

CRC Press
Taylor & Francis Group
6000 Broken Sound Parkway NW, Suite 300
Boca Raton, FL 33487-2742

© 1994 by Taylor & Francis Group, LLC
CRC Press is an imprint of Taylor & Francis Group, an Informa business

No claim to original U.S. Government works

ISBN 13: 978-0-7484-0113-0 (pbk)
ISBN 13: 978-0-7484-0112-3 (hbk)

Visit the Taylor & Francis Web site at
http://www.taylorandfrancis.com

and the CRC Press Web site at
http://www.crcpress.com

Library of Congress Cataloging-in-Publication Data

Catalog record is available from the Library of Congress

Contents

Foreword

From humble beginnings automating routine clerical tasks the computer industry is now attempting more daunting challenges; to support and expand the capabilities of human beings in more complex, innovative intellectual tasks. So far the results have not been very encouraging. Rather than liberating our creative capabilities, we often find that these empowerment tools are confining and inhibiting. A major reason is that we have too narrow a view of the human task that we are trying to support with our technology. If the undoubted power of modern technology is to be effectively harnessed, we need richer accounts of the sophisticated human skills we seek to empower.

Andy Dillon has sought a deeper understanding of the very sophisticated and significant skill of reading. His studies quickly dispel any notion that we laboriously read from the first page to the last. We can have many purposes when we pick up a text and we vary our reading strategy accordingly. The richness and subtlety of the reading skill that these studies reveal will give pause to any developer contemplating the opportunities of electronic texts.

As human scientists we always feel more comfortable when we can conduct our studies under controlled, experimental conditions. Andy shows the value of experimental studies but he also uses a wide range of other techniques. Each of them adds another valuable perspective to our understanding of reading. His work is testament to the importance of studying a phenomenon in context as well as in the laboratory.

These studies of reading were undertaken while Andy was a member of the HUSAT Research Institute. His work is an eloquent statement of the values and objectives of the Institute in seeking to ensure technological developments match human requirements and characteristics. It is a pleasure to acknowledge Andy's contribution to the intellectual development of the Institute and to commend this account of reading to everyone who would seek to understand it and to support it with technology.

Ken Eason
HUSAT Research Institute
November 1993

Acknowledgments

This book owes so much to my working experiences at the HUSAT Research Institute of Loughborough University of Technology in England where I spent many happy times between 1986 and 1993. HUSAT is a speial place as demonstrated by the fact that it has maintained an existence as a self-funding research and consultancy group since 1970 solely on the basis of the researchers' abilities in applying the methods and knowledge of the human sciences to the design of advanced technology. Long before the current vogue of marketing 'user-friendliness', HUSAT advocated usability and an appreciation of the users' and organizations' perspective in developing acceptable new technologies. More than this, HUSAT convinced the design and manufacturing worlds to support its work. While the 1980s produced many new research groups in this field, I know of no comparable institute worldwide.

In particular mention must be made of my closest working partners during these years, Dr Cliff McKnight, John Richardson, Dr Martin Maguire and Marian Sweeney who have put up with more than most, as well as David Davies and many others too numerous to mention. HUSAT also gave me a chance to work with two people who epitomize for me all that is positive in this domain: Prof. Brian Shackel and Prof. Ken Eason. More than either will probably ever realize they have had an impact on my thinking and convinced me that the concern for user-centredness in technological developments is an area worthy of intellectual and professional pursuit.

Time to write this book was partly funded by a visiting position at the Institute for the Study of Human Capabilities, Indiana University, Bloomington, USA where I met and argued with numerous people. Particular mention here must be made of Prof. Chuck Watson for developing my interest in individual differences, Prof. Tom Duffy of the Instructional Systems Technology Group and Prof. Tom Schwen's hit squad in the Centre for Media in Teaching.

1

The reading process and electronic text

People say that life's the thing, but I prefer reading
Logan Pearsall Smith, Art and Letters, *Afterthoughts* (1931)

1.1 Introduction

Books about information technology (IT) have a tendency to start with a
general statement about the continuing swift developments in the field of
computing and information processing and in so saying, the present author
has conformed to type. However, while it is typical to discuss such
developments in terms of falling hardware costs and technological advances, it
is, at least for the human scientists amongst us, more interesting to observe
such developments in terms of their influence on human activities. While
theorists talk of the information age, we are in practice creating an
information world where microprocessors interface between us and
innumerable as well as previously unimaginable activities.

The present book is concerned with one such development, the use of
information technology to support the activity known as 'reading' and in so
doing, to challenge the supremacy of paper as the most suitable medium of
text presentation. This area is receiving a lot of attention currently as hypertext
gives new substance to old ideas but such attention is typically directed more at
developing the technology than considering how and why it might be useful.
This book is concerned directly with the readers and thereby, the text designer
and author who must cater for them. In so being, it touches on numerous
related areas where documents of any kind play a part in human activities:
education, training, leisure, work and so forth. Its focus however is the users
or readers and the design of an information technology that suits them.[1]

1.2 The emergence of electronic text

For a medium that is so new perhaps it is surprising that a history of electronic
text can be even considered, never mind described. However, the idea of using

the electronic medium to support reading can be traced back several decades and no self-respecting writer on the subject of hypertext ever fails to mention such visionary thinkers as Bush (1945), Engelbart (1963) or Nelson (1987)[2] who in their own way advanced (and in some cases continue to do so) the concept of access to a world of knowledge through information technology. These thinkers paved the intellectual path to hypertext[3] and its underlying philosophy that humans should be able (from their desktop or laptop, at work, at home, or on the move) to locate, retrieve and use easily the store of human knowledge that lies in books, journals and associated materials in libraries the world over.

Despite its ancestry or philosophy, electronic text has had to wait for a technology to develop before such fantastic ideas could be embodied. The computer is that technology and only comparatively recent developments in microelectronics have enabled the concept of electronic text to be seen and not just heard. Feldman (1990) points out that despite the advocates of previous decades it is only in the 1980s that electronic text could really be said to have arrived. Prior to that it was conceived, talked about, and its potential imagined, but it did not truly exist.

Spring (1991, p. iii) puts it well when he states that:

> During the 1980s there was a little-noticed change in the world of computing. One day during that decade, I guess it was about 1985, more computing cycles were devoted to manipulating words than were devoted to manipulating numbers.

The precision of the time line is debatable but it is not difficult to see why the personal computer boom of the 1980s coupled with developments in digital information storage and presentation have made electronic text both feasible and culturally acceptable. Both these aspects are necessary for electronic documents to succeed. It is not enough that it can now be done, that electronic text can, for example, reduce the 20-volume Grolier Encyclopedia to a single compact disc (and still leave more than half of the disc free), but the world needs to be ready for electronic text. Readers must appreciate its relevance, its potential advantages and more importantly, they must want to use electronic text if it is to succeed.

While the information culture is emerging, the acceptance of electronic text currently lags behind its technical feasibility. As with most if not all technical innovations, the claims of its advocates wildly surpass the reality of its impact. With the exception of a small number of researchers, designers and keen amateurs, the idea of reading lengthy texts in their electronic as opposed to paper form tends to be viewed negatively rather than embraced whole-heartedly. Hence, most probably, you are reading this sentence through the well established medium of paper. It will take time and effort to identify the optimum forms for electronic text. Currently, designers lack the expertise and experiences that have evolved with paper text and the present work seeks to contribute to the increased research and development effort on this front. These are early days for the new medium (even if it is possible to distinguish between generations of electronic text) and one should avoid seeing electronic text as a competitor to paper in some form of 'either–or' challenge for supremacy. It is not inevitable that electronic text will replace paper as some writers have suggested (e.g. Jonassen, 1982) but it might displace it as more and more human activities become mediated by information technology. This should not be allowed to happen by accident though, as a side-effect of increased computerization, we must seek actively to influence the process for the better so that the positive aspects of electronic text are accentuated. To achieve this psychologists, information scientists, sociologists, writers and (most importantly) readers must influence the design process. The history of electronic text, and of information technology in general, is still being written and will be shaped by many forces.

1.3 The aims of the book

The major aim of the present work is to examine and subsequently to describe the reading process from a perspective that sheds light on the potential for information technology to support that process. Current research suggests that paper is by far the preferred medium for reading though there is less consensus on why this is, in fact, the case. It is clear that simply transferring paper formats to the electronic medium is insufficient and often detrimental to use. Therefore clarification of the future role for electronic text in our information world would seem to be an issue worthy of investigation. From a psychologist's viewpoint, the issue is far less to do with technical feasibility – can it be built? (although of necessity at times this question rears its head) – than with how human cognition and behaviour places constraints on, or provides clues to, the usability (and hence the acceptability) of current and future technologies.

This however, is not a psychology text and it most certainly is not a text on the psychology of reading as that descriptor is usually understood; it is a book about reading and the examination of that process from a primarily psychological perspective. If it needs classification it could be described as an ergonomics or human factors[4] book but even then it may not conform to others' expectations of that subject's style. If classification proves awkward then so much the worse for the classification. The design of usable electronic text and the study of human information usage are issues that deserve direct attention and as will be discussed frequently throughout this book, rigid disciplinary boundaries are not helpful here.

Traditionally, ergonomics (or human factors) has offered itself to design engineers as an evaluative discipline, equipped with the tools and methods to assess the performance of human operators with developed systems. In recent years, as a result of the need for more rapid design cycles and increased competition amongst developers, a need for earlier inputs to the product life cycle has arisen. Such inputs, in the form of models, guidelines, checklists and design tools attempt to package ergonomic knowledge in a form suitable for engineers to consume and apply. This has not proved an easy task and there are many in the human factors discipline uncomfortable with this role (not least the present author).

This book will not concentrate on the issue of technology transfer between ergonomics and design but is aware of its existence as a yardstick against which, rightly or wrongly, the value of current human factors work is often measured. Consequently, such issues cannot be avoided in contemporary ergonomics and a second aim of this book is to develop a framework for considering user issues that is potentially applicable to the earliest stages of electronic text design. Though it should be emphasized here and now, it is not expected that such a framework could just be handed over to non-human scientists and flawlessly applied by them as so often appears to be the intention of HCI professionals. The emphasis throughout the work is therefore less on empirical investigations of various user interface variables (though these are to some extent present) and more on identifying the primary human factors underlying document usage through knowledge elicitation techniques and observation of usage patterns, with a view to forming these into a conceptual framework that can be represented in a form relevant to design.

1.4 The scope of the book

In simple terms, this work is concerned with the human as reader, i.e. user of textual information. However, its remit is broad by comparison to much of the theoretical psychological work in this area which tends to define reading narrowly as the transformation of visual images to perceived words or the extraction of meaning from structured prose. Such positions, though extremist, are both tenable and frequently published. Indeed much of the work in experimental psychology on reading assumes one or other interpretation; see for example, Just and Carpenter's (1980) model of reading which includes comprehension and Crowder's (1982) description of where reading begins and ends which explicitly excludes it. Instead, this book covers the range of issues involved in using such material, from identifying a need for it, locating and selecting it, manipulating it to ultimately processing it. Therefore, while its interests are primarily psychological, the consideration of alternative perspectives from disciplines such as information science, computer science and typography are both necessary and insightful.

In the present context therefore 'reading' implies situations where the human will engage the medium to perform any of the range of activities typically afforded this descriptor in common parlance. Thus it covers a variety of scenarios ranging from proof-reading a document to examining the contents of a book to see if it contains anything of interest, but omits those that have reading as a component but necessarily secondary part, such as text-editing. Furthermore, included under this term are the associated activities of location and manipulation of textual information that invariably precede and are concurrent with these tasks in the real world. How far one extends this is a matter of common sense. Obviously walking into a library or bookshop necessarily precedes the act of reading a text there but should not be considered part of the act of reading itself. However, within the broad reading task scenario, searching for a book or browsing the spines of numerous journals in order to locate a specific edition are part of reading in the sense used here.

By text is meant any document, with or without graphics, that can be presented to a reader as an information source. Thus it includes those documents that we are typically exposed to in everyday life such as newspapers, books, magazines, technical manuals, letters and so forth, as well as less traditional texts such as electronic documents and textual databases. Though termed text, this descriptor might include those documents that have a large graphical content, such as catalogues, but not those that are primarily graphical such that they relegate alphanumeric text strings to secondary importance, e.g. maps. That said, much of what follows in terms of usage and structure of information would seem relevant to the design of even highly graphical information displays.

The term 'electronic text' is used by means of contrast with paper documentation, i.e. it is any text that is presented on a computer screen. For the purposes of this book the terms electronic and screen-presented text are used synonymously and imply presentation via computer screens. They do not refer to any other form of screen-presented text such as microfiche, microfilm or slides which involve magnification and projection rather than electronic processing. The term includes hypertext and non-hypertext. Like its paper equivalent, electronic text may contain graphics. However, it does not cover the term hypermedia which is often mistakenly assumed to be simply hypertext with extra graphics.[5] Generically, the term 'information space' is used to cover all published materials: text, hypertext and hypermedia. Where it is employed in this book its meaning is implied by the context of use unless otherwise stated.

Obviously it is impossible in the present situation to cover all manifestations of reading texts as the terms are defined here and indeed the book concentrates primarily (though not exclusively) on scientific literature such as academic journal articles or technical manuals for its empirical investigations. However, even academic articles are lengthy texts which, it will be shown, are read in a variety of ways that extend their comparability with other document forms. Similarly, the use of both software manuals and booklets also reported here broaden the coverage of the work. Thus the issues raised and concepts presented in the final framework are intended to be generic and applicable to most text forms and reading situations in as much as electronic media might influence their interaction.

1.5 A note on methods

This work, by choice, avoids many of the issues of learning to use innovative technology which some would see as a natural role for a human factors researcher (and the subject matter of numerous workers who claim in fact to be examining usability as if the latter concept was nothing more than ease of learning). It is not that such research is seen as irrelevant but that the author believes that well-designed systems should start from a premise of supporting specific users performing certain tasks rather than worrying prematurely about ease of learning. In this application domain, design is necessarily speculative; there are few, if any, rules or established systems to react to or design against. Contrast this, for example, with designing a new text editor where not only does a large body of knowledge on how users perform such tasks exist but designers can examine numerous existing products to inform their own design. Consequently, the author sees the role of psychology and other human sciences in this area as a dual one of guidance and suggestion, using knowledge of human cognition and performance to constrain the number of potential design options while informing speculation on how humans might like things to be. Such work necessarily precedes learnability research.

The stated aims and approach of this work have dictated the methods employed, not only in this book but also in my professional work. This book is an applied work, a study of human factors carried out during the development and evaluation of real products. In order to identify how electronic text systems are designed and the best role for human factors knowledge in this process, it is necessary to involve oneself in the process, to be part of a design team, to develop electronic texts and to assess the consequences of one's work. Only in this way can one really appreciate what is needed, what questions arise, what type of human factors input is useful and what are the limitations of the discipline's (and one's own) knowledge. Theorizing from without may have proved intellectually stimulating but would have been insufficient. To paraphrase Card *et al.* (1983), 'design is where the action is' in human–computer interaction (HCI).

On the face of it, the human scientist would appear not particularly well armed for action of this kind. The traditional strengths of psychology and ergonomics lie in designing and conducting formal experiments, planning work in detail before carrying it out, controlling for all undesirable sources of variance, and reporting the results in a conventional academic form. As a result of such an approach, a substantial literature has emerged on the presumed usability of various interface features or the significant problems associated with certain products. Essential as such work is in building up the bedrock of empirical knowledge, on its own it cannot provide the answers to many of the questions posed here.

Examining the issue more deeply however, one might come to appreciate that the human scientist is the ideal person to become involved in the design of interactive products. Equipped with knowledge of human behaviour and dispositions, skilled in the consideration of how certain design features influence performance, the human scientist can make the distinction between popular conceptions of users based on opinion and myth, and accurate models based on reasoned argument and psychological findings. One should be able to distinguish between occasions when approximate answers will suffice and when only formal experimental evaluation will provide answers. Most importantly, one should be able to identify gaps in the knowledge base of design that only the human sciences can fill or hope to fill. In short, the human scientist may be seen as the only suitable candidate for the job.

This is the philosophy of the present work. Involvement has been achieved by working at the HUSAT Research Institute in Loughborough,[6] the largest university-based research and consultancy institute in Europe dedicated to the application of the tools and techniques of the human sciences in technology design. After joining HUSAT in 1986, I became a member of a research team specifically investigating electronic text design a year later. In the subsequent period this team has worked (and in part continues to work) on four long-term research projects related to electronic documentation and human information usage. In the first, Project Quartet (Tuck *et al.*, 1990) the present author was a member of a three-man team of psychologists at HUSAT who worked with other research teams at three universities consisting, in the main, of computer scientists. The goal of this work was to investigate the impact of information technology on scholarly communication. In the other two – Project OCLC (McKnight *et al.*, 1988) and Project CHIRO (McKnight *et al.*, 1990b) – the author was a member of the same team at HUSAT investigating both the interface issues associated with access, delivery and usage of lengthy electronic texts from remote sources, and the use of hypertext as a collaborative work environment for scientists. The fourth – Project HyperLIB – continues at this time, investigating the potential for hypertext to support greater exploitation of information stored in academic libraries. These projects, coupled with a variety of short-term consultancies on human factors for numerous industrial software companies and departments throughout Europe, were, and in some senses continue to be, the testing grounds for many of the ideas outlined in this work.

The nature of the work and the impact of the author's findings on real-world applications means that most of the studies reported here were not laboratory exercises isolated from practical concerns, but investigations carried out during design processes to provide answers to genuine questions, to resolve design issues or to test specific design instantiations. It is an example of the psychologist as applied scientist, part designer, part team member and part user, all the while monitoring his own and his colleagues' work in a meta-analytic fashion. It is my contention that such a process is not only a worthwhile method of research but also the only sure way for suitable knowledge of this field to be gained.

Generally, the techniques and methods employed in this book vary from the experimental to the exploratory. The use of a method was determined by the type of information sought – what needed to be known led to the choice of investigative methodology. Expertise in or familiarity with a technique was never considered sufficient justification for its employment. For example, at the outset it became clear that information on how readers view texts and their interactions with them was in short supply in the literature. This gap in the knowledge base is in part due to the inherent difficulties in capturing such information in a valid form. Experimental techniques are impracticable in such situations and reliable questionnaires on such matters do not exist yet. In order to overcome this, information was gathered employing a mix of knowledge elicitation techniques from the 'harder' (in the comparative sense only) or more objective such as repertory grid analysis, to the 'softer' or more subjective ones such as interviewing until satisfactory answers were obtained. Binder (1964, p. 294), a psychologist and statistician gave inspiration; he wrote:

> We must use all available weapons of attack, face our problems realistically and not retreat to the land of fashionable sterility, learn to sweat over our data with an admixture of judgement and intuitive rumination, and accept the usefulness of particular data even when the level of analysis for them is markedly below that for other data in the empirical area.

I have provided the sweat and hopefully the correct 'admixture of judgement and intuitive rumination', and it is for the reader alone to determine whether the results have achieved the appropriate status of usefulness.

1.6 Outline of the book

The book commences with an examination of the concept of usability as it is applied to the design of interactive technologies and the potential application of human factors or ergonomic knowledge and techniques in this domain. Following this is a thorough review of the experimental literature on reading from paper and screens. This review is divided into three major parts. The first describes the reported differences between reading from paper and from screens. The second part reviews the analyses of these differences that have resulted which can be seen as an attempt to identify the basic differences between the media and subsequently isolate the crucial variables in terms of three levels: perceptual, motor and cognitive issues. The third part of the literature review highlights potential shortcomings in the comparative feature-based approach and raises issues relating to text and task variables.

Chapter 4 concentrates on the value of the existing human factors literature to electronic document designers, highlighting the underlying acceptance of uni-disciplinary (primarily cognitive) views of the reading process manifest in the literature. The problems of applying findings from this work to the design and evaluation of an electronic text system are highlighted by reference to a case study carried out by the author.

In order to overcome some of the perceived shortcomings of previous work, particularly in relation to reading context, an analysis of the categorization of texts, including readers' views of text classifications, is presented in Chapter 5. This work emphasizes the world of texts as partly given and acts as a stimulus to more detailed examination of readers' interactions with two distinct text types: academic journals and software manuals in Chapter 6. In combination these data shed light on what readers see as important in text usage and how electronic versions might be designed to convey benefits. Chapter 7 considers further the literature on readers' impressions of structure and shape in information space and how this might relate to the much discussed potential for navigation problems with electronic information.

Following this work, a framework is proposed in Chapter 8 which represents the ergonomic factors involved in using a text and suggests the variables to consider in designing an electronic document. The framework consists of four interactive elements which reflect the issues dominating the reader's attention at various stages of the reading process. The interactions between these elements are described and the role of such frameworks in human factors is assessed. This framework is then used to derive predictions of reader performance in certain tasks, and these are tested in two studies using hypertext and paper. Verbal protocols from one study are employed to examine the relevance of the elements described in the framework.

The final chapter reviews the work of the book in the light of its stated aims and highlights areas for future research. A sequence of human factors inputs to the design stage which should aid usability of the resultant application is made explicit in this chapter. The final section of the book assesses the realistic prospects for electronic text in the information world that is emerging.

Unlike most hypertext advocates I will not encourage you to jump all over the book – I have written it this way in order to highlight how the ideas developed and are related. However, since experimental evidence has demonstrated repeatedly to me that educated readers rarely read a book in a completely linear fashion anyway, I should neither encourage you (since you need no encouragement) nor discourage you (because it is likely you are reading this section too late for it to make a difference!).

Notes

1. Throughout the text, footnotes are used to examine related points or inform interested readers on alternatives and caveats. The fact that you are reading this one has probably demonstrated to you that they are not the most usable literary device, particularly where publishers make authors place them off the page of occurrence. While I shall keep footnotes to a minimum in the book, I encourage you to read them as you come across them for maximum appreciation. In so doing, you might become acutely aware of one possible advantage for electronic text.
2. The comparative recency of this reference stems from the fact that much of Nelson's work is self-published and/or distributed. His ideas actually gained currency far earlier than this date suggests.
3. Hypertext is an electronic form of text presentation that supports the linking of nodes or chunks of text in any order. It has been defined and explained so often in the previous five years that I will not attempt to explain it further. If any reader finds this description insufficient or really does not know what the term implies, a quick read of any one of a dozen introductory texts on the subject will provide the answer (e.g. McKnight *et al.*, 1991).
4. Ergonomics is the scientific study of the person in relation to the designed environment (usually work). In the USA, ergonomics was more frequently known as

'human factors' until recently when a move towards using the more European term 'ergonomics' has been observed. The terms will be used interchangeably throughout this book.

5. These terms are often used synonymously in the literature which is erroneous. Hypermedia implies the use of multiple media to present a range of information such as text, sound and animation, not just the marriage of text and graphics on a single medium. Hypermedia information sources have no paper equivalent.

6. HUSAT stands for Human Sciences and Advanced Technology.

2

Electronic documents as usable artefacts

Sirs, I have tested your machine. It adds a new terror to life and makes death a long felt want.

Sir Herbert Beerbohm Tree on examining a gramophone player.

Hesketh Pearson *Biography* (1956)

2.1 The emergence of usability as a design issue

Electronic documents are one more application of an information technology that has proved immensely successful in supporting tasks as diverse as flying an aircraft and handling financial transactions. The computer has expanded beyond its initial remit as a number cruncher for scientists and become a storer, manipulator and presenter (and even a supposedly 'intelligent' analyser) of all kinds of information. What really could be more appropriate for a so-called 'information technology' than to store, manipulate and present text for readers?

Before examining issues of direct concern to electronic documents, it is worth considering their emergence from the broader technological perspective that became known as the 'information revolution'. Without wishing to delve deeply into the history of computing technology, it is important to appreciate that computers moved from expensive, large and dedicated systems used only by specialists in scientific or business domains in the 1950s and 1960s, through a series of stages involving ever smaller and cheaper instantiations, operated by less specialist users, employing a variety of higher level languages in the 1970s. The shift from the bulky mainframe to the desktop computer late in that decade paved the way for widespread usage by people not particularly well versed in programming languages, operating systems and hardware issues.

As technological developments in the computing field advanced, it became clear that cost or functionality were no longer the major obstacles to massive uptake in society. Falling comparative hardware and software costs meant that computing technology became affordable for many applications. However, bottlenecks in the technology's acceptance appeared largely as a result of problems associated with users of the new systems.

Originally, the users of this technology were experts in its design and inner workings, often building and maintaining it themselves and certainly trained and knowledgeable in the effective operation of their computers. Often this meant spending several years learning to write and read specialized programming languages in order to 'instruct' the machine. However, as early as the late 1960s some computer scientists (as they became known) noted large variances in the time it took programmers to produce the same codes and that different end-users responded differently (and unpredictably) to the finished design (e.g. Weinberg, 1971). It started to become clear that human costs were also important in the development and particularly the usage of this revolutionary technology. As the new waves of computing applications emerged, it was human resistance that came to be seen as holding up the new technology's exploitation.

For any technology to reach a wide audience of potential users it must minimize the amount of specialist knowledge or training required to make use of it, otherwise it is destined to appeal only to trained or specialist user groups. The rise of the 'casual user', persons with little or no specialist training or knowledge – noted informally by many early in the 1970s but formally named first by Cuff (1980) – was an obstacle to uptake that could only be crossed if the technology was made suitably easy to use by its developers. As a result, in the 1970s a research effort started in earnest that carries on to this day. This research took as its subject matter the process of human–computer interaction with a view to identifying how computing artefacts could be designed to satisfy functional requirements while minimizing users' difficulties. The field became known by the same name as its subject matter – human–computer interaction or HCI[1] – but more accurately should be described as an extension or application domain of the existing discipline of ergonomics or human factors. In fact, ergonomic concern with computer interfaces can be traced as far back as the work of Shackel (1959), considered by many to be the first published paper on HCI.

HCI focused attention on the interface between the person and the machine such that interface design has now become a central concern to interactive product developers. The interface is essentially a communication channel that affords the transfer of modality independent information between the user and the computer. Possessing both physical (e.g. keyboard, mouse, screen, etc.) and representational qualities (e.g. being designed to look like a desktop, a spreadsheet or even a book), the user interface is for many people, all they see and know about the computer. Immense research effort has been expended on studying how this transfer of information can be best supported so that users can exploit the undoubted power of the computer to perform tasks. Certainly progress has been made and contemporary interfaces are often (but not always) far superior to typical user interfaces of even a decade ago. Yet problems still exist and user error, resistance and dissatisfaction often still result from poorly designed technology. The field of HCI still has much work to do to ensure user issues are adequately addressed in design.

This brief introduction summarizes significant developments in recent design, technological, industrial and social history that deserve entire books to themselves to cover appropriately (e.g. Forester, 1985). However, it is against this backdrop that we must examine the central issue in the design of electronic documents – their usability.

2.2 Usability as part of the product acceptability equation

The success of any interactive product or system is ultimately dependent on it providing the right facilities for the task at hand in such a way as they can be effectively used, at an appropriate price. If it can satisfy these criteria then it can aptly be described as acceptable.

Shackel (1991) discusses product acceptability as an equation involving the relationship between functionality, usability and cost. Functionality refers to the complete range of facilities offered by a tool or product. Basically, any technology offers its users the potential to perform a variety of actions. Some products offer more functions than others and in certain product domains extra functionality is often used as a major selling point in marketing the technology. Obviously for a product to be acceptable it must offer users the range of functions they need.

However, merely providing the facilities to perform a task serves little purpose on its own if the user cannot discover them, fails to employ them effectively through error or misunderstanding or dislikes them sufficiently to avoid usage. Such problems are not functionality issues but usability issues and it is these that are determined largely by the user interface. It is not enough that a system is efficient, cheap and highly functional in task terms. If users have problems with it then such features are never exploited and the product must be deemed at least a partial failure.

This point can easily be appreciated by considering the typical office telephone. Many telephone sets offer the potential to transfer calls to colleagues, to form conference calls, to answer a telephone on another desk that is linked in a network, or to place call-back commands on engaged lines. These are all functions. However, many users are incapable of exploiting these functions for a variety of reasons, e.g. they were never trained in using the telephone, they cannot remember the codes required, they do not want to read the accompanying manual (if they can still find it). The keypad on the telephone itself rarely affords sufficient clues and a user is therefore required to remember e.g. that **3 might answer a colleague's ringing telephone. These issues are a reflection of the interface design of the telephone set and therefore provide a statement of its usability. Such a design clearly reflects an emphasis on functionality at the expense of usability. Usability is an increasingly popular concept that is at the heart of user-centred system design, a philosophy of design that will be outlined later in this chapter and recurs as a theme throughout this book. Varyingly equated with such concepts as 'user-friendliness'[2] (Stevens, 1983) or 'ease of use', usability is often employed without formal definition. However Shackel (1991, p. 24) states that the usability of a system can be defined as:

> [its] capability in human functional terms to be used easily and effectively by the specified range of users, given specified training and support, to fulfil specified range of tasks, within the specified range of environmental scenarios.

The key aspects are 'easily' (i.e. the user – with or without training as specified – must not find it difficult to utilize the functionality of the system to perform a task) and 'effectively' (i.e. performance must be of a suitably high level, however defined). Many systems could satisfy one of these criteria and either be easy to operate but not very effective in terms of task performance, or be extremely difficult to use but, if ever mastered, could support very effective task performance. Neither of these would be deemed very usable in terms of the definition above.

The International Standards Organization (ISO) is introducing a new standard for information technology (ISO 9241) that covers the issue of software usability and uses a slightly modified definition which equates usability with aspects such as effectiveness, efficiency and satisfaction of use in specified contexts. The emergence of usability at the level of international standards for design is testimony to the growing importance of such ergonomic issues in technology production.

Given the concept's definition, the issue of context is obviously critical to an understanding of usability. It makes no sense to describe a tool or technology as usable or unusable in itself. Any tool is made for use by certain users, performing particular tasks in specific environments. Its usability can only be evaluated meaningfully in relation to these contextual variables. There is no usability metric that can be employed independently of these factors since it is their dynamic interplay that determines efficiency, effectiveness of use and satisfaction etc. Thus any statement of a product's usability must be made in relation to the user, task and environmental types for which it is designed. One must therefore view with suspicion any claim by a design company to have developed 'the most usable product' of its kind if the accompanying contextual details are not provided. A more correct statement would be, e.g. 'the most usable product for able-bodied data-entry clerks working in a typical office environment' (although even this statement leaves much information unspecified). Unfortunately, such lengthy descriptors do not make convenient advertising slogans and it is likely that appeal to 'ease of use' in marketing campaigns will continue to flourish.

The contextual aspect of usability also means that it is incorrect to assume that, by virtue of reference to the implicit ease-of-use issue, all usable products must necessarily be simple to operate. Some technologies are only designed for specialist use and they may be applicable only in complex task domains. In such scenarios, usability for specialists will almost certainly not result in a product that is simple to use for anyone else. Indeed, the product might even be totally unusable by such non-specialists precisely because it has been designed to be usable in other contexts. This is also the case with learnability. It is wrong to assume that all usable technologies must necessarily be simple to learn to use. Some tasks require such a level of skill and knowledge to perform – open-heart surgery comes to mind! – that it would not necessarily mean an instrument the expert user found usable (in this case the surgeon) must be easy for anyone else to learn to use effectively and satisfactorily. This is a point that is rarely grasped by those concerned simply with putative ease of use and its assumed dependence on the presence of certain features in the human–computer interface such as windows, graphics and menus. It is possibly such misunderstanding of the real meaning of the concept that underlies some of the ergonomics profession's criticisms of it.

The concept of usability is sufficiently broad for it to cover many aspects of contemporary research in human–computer interaction (HCI). One distinct strand involves the formalization of its definition and its evaluation methods. Another is the analysis of what attributes of an artefact affect its usability. It is important to distinguish these as their merging under one heading is likely to blur important conceptual and methodological distinctions and lead to confusion in discussion of usability and the human impact of technology. These distinctions are briefly explored in the following sections.

2.3 Usability evaluation

It is generally accepted now that interactive technologies cannot be designed from first principles so as to guarantee their usability and that systems need to be tested in some way. The term 'usability evaluation' covers a diverse range of activities all concerned with assessing the likely user response to a proposed design. According to Sweeney *et al.* (1993) usability evaluations generally take one of three forms: theory-based, expert-based and user-based.

Theory-based evaluations generally involve the application of analytic models of user performance to design models or system specifications. In effect, such models are articulated in a form that enables developers and designers to gain an impression of likely user response to a proposed design and thereby assess that design's suitability to support task performance. The most successful such HCI formalism has been the GOMS family of models (Card *et al.*, 1983) which support the prediction of expert users' task-completion times based on an analysis of ideal actions (cast in terms of information processing speed in a simplified cognitive model of information processing stages). Shortcomings in application exist (e.g. the models may be of little use in the analysis of complex tasks involving much user discretion in performance) and the need for expertise in their derivation has generally limited formal models' use. But for early input to the design process of specific tools these formalisms can be of value. To date however, the design world has been relatively slow to pick up on them and their commercial viability remains unproved. This issue will be dealt with further in Chapter 8.

Expert-based evaluations rely on trained human factors professionals examining a design in order to identify likely usability problems. These evaluations can range from high-level 'quick-and-dirty' examinations to structured 'walk-throughs' where the evaluator examines the design in search of strict compliance with specified standards or target guidelines. An advantage of this approach is its comparative low cost. In theory, one ergonomics expert should be able to evaluate a design and quickly identify any problems. However, this view is difficult to substantiate on the basis of evidence.

Nielsen (1992) for example, a major proponent of the cognitive walk-through or heuristic evaluation approach, reported comparative data which showed that while usability specialists were better than non-specialists (those who had received a short training course), it still took three or four experts to identify over 75 per cent of known usability problems in a test product. Furthermore, evaluators with detailed knowledge of the task domain in which the product would be located, proved better than the regular evaluation specialists. Wharton *et al.* (1992) report problems with the standard cognitive walk-through approach to evaluation and conclude it is unsuitable for commercial use without substantial modification. In particular they stress the value of involving end-users in design evaluation and more than one specialist evaluator. Even then, they state that one should expect to miss some possible usability problems.

By involving users in the process one enters the user-based evaluation domain. Traditional psychology-based procedures emphasize the role of experimental trials with typical users to gauge usability. In essence it is like running a simple psychological experiment in which typical users are exposed to the product and asked to perform a selection of typical tasks. Their interactions are recorded, often on videotape, and the evaluators use this data to rate the product's usability and identify weaknesses in the interface. However, the cost involved in performing such evaluations and the need for formal training in experimental design means that design companies are often slow to embrace this approach fully. Furthermore, such evaluations require a reasonably finished or high-fidelity version of the product to test and this means they can often only occur late in the design cycle where the opportunities to alter the design in any significant fashion are limited by cost considerations.

Karat *et al.* (1992) report a comparison of such evaluation methods which indicated that user trials are better than design team walk-throughs which are in turn better than individual expert walk-through. Their estimates suggest that there is only 33 per cent commonality between the approaches, i.e. only 1 in 3 potential usability problems is identified by all methods. The data thus suggest that user trials form the most reliable method but that each has particular strengths that support its use for certain products and at certain stages of the design cycle. As always, the quality of the evaluator is a crucial variable in the success or value of any usability text.

2.4 The laboratory versus the field

Another issue in evaluation concerns the use of laboratories and field sites for testing usability. In the 1980s many companies, in a highly-publicized attempt to display a firm commitment to usability, invested heavily in usability laboratories to test their products (Dillon, 1988b). In effect, these laboratories resulted in the generation of a lot of videotape of users interacting with tools in simulated offices but in the main served only to highlight the shortcomings of evaluation methodology rather than the advance of usability engineering. In fact, a major weakness of the user-centred design philosophy has been the ease with which one can claim adherence to it without a commensurate shift in the design process.

As laboratory studies have often failed to predict real-world usability, a move towards taking the contextual nature of usability to its limits has emerged and some proponents claim that usability can only be assessed in the field. This perspective is supported by research that shows how some laboratory evaluations have failed to pick up usability problems that have been found in the field (e.g. Bailey *et al.*, 1988). However, while the ultimate proof of a design is its reception in the real world as demonstrated over time, field tests, like laboratory trials, are only as good as the evaluator who plans and performs them. Such findings say less about the problems of laboratory evaluations than it does about the assumptions these evaluators made on tasks performed with the tool which were subsequently shown to be false in the field. As has been stated, evaluations must occur in context and if the context can be reproduced accurately in a laboratory evaluation than all well and good.

2.5 What makes a technology more or less usable?

The process of evaluation is important but in itself tells the design team little about how to improve usability or design it into the system at the earliest stages. Making the leap from evaluation result to design attribute that influences usability requires knowledge of the psychology of HCI, the literature on interface design in general, and specific knowledge of the application domain in which the designed artefact is located. Relevant psychological characteristics of the interaction process include the information processing characteristics and limitations of the user such as the tendency to interpret information in terms of pre-existing knowledge structures or schemata, the selective nature of attention, the nature of visual, auditory and haptic perception, the limitations of short-term memory and so forth. It is not the intention of the present book to cover material such as this which is dealt with at a range of levels and breadths in a vast existing literature, e.g. Gardiner and Christie (1987) offer a good introduction to HCI related psychology.

The literature on interface design can be traced back over twenty years and in that time innumerable interface features have been put to empirical evaluation and comparison. In one sense, this work reflects a flawed research paradigm that sought to identify the core set of 'good' interface features that could be used as building blocks for interface design. However, this is an extreme view and few ergonomists or human factors professionals carrying out such comparisons really believe such a 'recipe-book' approach is now feasible, not least because of the situated or contextual nature of human–computer interaction. Nevertheless, the literature contains many guidelines based on work which purports to show that say, menus are better than command languages, large windows are better than small windows or the mouse is better than function keys for input, much of which is extremely useful in designing user interfaces but only if interpreted appropriately. The relevant aspects of this literature for the designer of electronic texts will be critically evaluated in the next chapter. Now, it is important to place this discussion of usability within the wider process of systems design.

2.6 Towards effective user-centred design processes

Traditional models of the design process make reference to numerous stages such as conceptual design, specification, designing, building, and testing, through which a product is seen to advance logically. In software design, structured decomposition was the dominant paradigm as manifest for example in the waterfall model – so called since activities are supposed to be completed at one stage before the next stage commences, and there is no need (or scope) to move backwards in the process. In the context of such processes, the traditional role of ergonomics professionals has been to develop suitable test procedures for the evaluation stage as briefly outlined above. Given what has been said of usability thus far, it is easy to see the relevance of the ergonomic approach to that stage. However, this has long been a source of complaint by ergonomists who have rightly pointed out that by the time they receive a product for evaluation there is little chance that they can persuade a company to make drastic alterations despite sometimes strong evidence of usability problems. Furthermore, problems identified as this stage could often be traced back to faults that ergonomists claimed they would have been able to identify far earlier in the process if only they had been consulted then. However, if one cannot accurately prescribe usable interfaces in advance (which ergonomics cannot) then a problem exists in terms of the role ergonomists can expect to play at the pre-evaluation stages of the design process. Being seen as evaluators of (other) designers' work, and consulted only when others see ergonomics as having something to offer, is guaranteed to limit the profession's input.

In effect, ergonomics has sought to resolve this dilemma by advocating an alternative process for artefact design that is generically termed 'user-centred' (e.g. Norman and Draper, 1986; Eason, 1988). User-centred design advocates a process that will increase the chances of an acceptable product being produced even if in advance it cannot be competely specified. The general philosophy is one of design–test iteration within the product life cycle, in contrast to more traditional phased design models which emphasize the logical progression of products from specification to design stages, to building and finally to testing.

In reality, design is rarely the smooth linear sequence of developmental stages many models suggest (except maybe in the eyes of senior project managers or financiers), rather it is an iterative sequence of events consisting of many sub-cycles and complex, often poorly understood and certainly opportunistic processes involving aspects of all stages. It is precisely this cyclical and opportunistic character that user-centred design methods seek to exploit. The fundamental premise of user-centred approaches (as the name suggests) is the addressing of user-relevant issues at all stages in whatever form is appropriate. Thus, assuming a generic representation of the design process as the eventual progression from specification, conceptual design, physical design, evaluation and release (with apologies to advocates of other 'key' stages) then the user-centred approach is to find a means of addressing relevant user issues at each and every stage.

Proponents of user-centred design suggest that it is insufficient just to wait for a usability evaluation late in the design process to gain feedback on user issues. Instead they argue that a system's eventual success is largely related to the degree to which user issues are addressed early in the design process. It is easy enough to see how standard usability evaluation processes can be applied as soon as some representation of the product is available to be tested. With the advent of cheap and fast prototyping applications it is now possible to mock-up a design for an interface rapidly and then test it and redesign accordingly thereby effectively removing the problem of only influencing the design process when it is too late to change anything. Little more needs to be said of this. Rapid prototyping is a major step forward and it should be simple enough to see how all that has been said so far in this chapter about usability can be applied to prototypes. However, there is more to user-centred design than just bringing the standard evaluation role forward in the product life cycle and many ergonomists are seeking to impact the very earliest stages of design that lead to the specification of the product prior to any prototype of it.

Theoretically and practically, earlier involvement is necessary since the ability to produce a cheap and quick model or prototype of the system is not itself logically sufficient to produce usable designs. Prototypes might continually be built and tested but not approach acceptable levels of usability if information about the target users and their tasks is weak or incorrect. Thus, according to user-centred principles, ergonomics needs to have impact at the stage prior to any building of the product, i.e. the requirements and conceptual design stages. Here, ergonomics seek to influence the very conception of the design and how it might look in accordance with informed views of the target users and their tasks.[3]

One important human factors input for the earliest stage of design is user requirements analysis. This can involve a set of procedures such as stakeholder identification, and user and task analyses, the basic aim of which is the provision of as much relevant, accurate information as possible about user-related issues that are likely to influence the eventual usability of a system. This information might be general or specific but it should act to constrain the range of design alternatives to consider and identify any crucial shortcomings in the subsequently proposed system. In this sense, the first prototypes are influenced for the better and the number of design–test iterations can be reduced.

2.6.1 Stakeholder identification

Stakeholders are all the key people who have an interest in the system to be designed. Thus it includes not only the target users but also those designing and maintaining the system, those offering user-support such as trainers and system managers, and those whose work might be affected indirectly by the system, e.g. recipients of output. This is particularly important in organizations where firm boundaries between direct and indirect users of a system and its outputs are blurred. For some applications the target end-users are the only key people involved, for others the stakeholders are many.

Each stakeholder should have the various events that could occur in their use of the system identified. This is termed 'scenario identification'. It naturally emerges for direct end-users during task analysis but is often difficult to predict in advance for all stakeholders in the technology. Gardener and McKenzie (1988) recommend considering scenarios in terms of criticality, difficulty, frequency of occurrence and any other distinctions viewed as relevant to that application. Ideally all scenarios should be explicitly considered in the design specification but this is invariably impossible. However, the process can be used to identify the critical ones to address for each stakeholder.

2.6.2 User analysis

Once stakeholders have been identified, key stakeholders should be analysed in detail and among the key stakeholders will always be the users. User analysis involves identifying the target end-users of the system and describing them in as much detail as possible in terms relevant to system usage. This might consist of any particular skills they have or need to be trained to acquire and the possible constraints these might impose on system effectiveness. It might also include identification of particular constraints imposed by working/usage environments such as noise and light conditions, which though not strictly user characteristics might often interact sufficiently with user performance to be considered a relevant factor.[4]

In the early years of human factors work on computer design various attempts were made to classify users into types in the hope that an acceptable classification system would enable accurate predictions about usability to be made at the outset of design. Unfortunately, many of these failed to demonstrate any validity and currently few typologies based on individual psychological differences (e.g. personality, cognitive style, etc.) can be shown to have any relevance to the earliest stages of software design. However, in broad terms, user analysis has utility in making explicit any assumptions about end-users and their characteristics.

2.6.3 Task analysis

Task analysis may be defined as the process of identifying and describing units of work and analysing the human and equipment/environmental resources necessary for successful work performance. It is often used synonymously with task description, job analysis, work study, etc. which is not accurate as each of these terms refers to techniques that share some aspects of task analysis but ultimately include issues that are insufficient for, or distinct from, pure task analysis. In the present context, the term is taken to imply the analysis, synthesis, interpretation and evaluation of task requirements in the light of knowledge and theory about human characteristics.

Historically, numerous techniques for task analysis have been described (e.g. Miller, 1967; Annett and Duncan, 1967; Moran, 1981). Broad distinctions can be drawn between them in terms of level of detail addressed (the 'grain of analysis'), formality of procedure involved and expertise required by the analyst. The basic principle common to all such techniques however, is the identification of procedures performed by the human during the task and the subsequent decomposition of these into their constituent cognitive, perceptual and physical elements. Psychological knowledge is brought to bear on the analysis and inform on optimal organization and synthesis of the cognitive and perceptual elements while physiological knowledge is related to the physical elements where relevant. With software, the cognitive ergonomic issues are primary, physiological issues are mainly related to the hardware side of information technology. There is not always such a convenient separation. For example, input devices are essentially a hardware issue for which cognitive aspects are sometimes discussed.

In the HCI domain new analysis techniques are continually emerging to aid designers of interactive computer systems. Wilson *et al*. (1986) review 11 such techniques but Diaper (1990) comments that this is probably an underestimate of (by then) current procedures. Despite the advantages of task analysis at the specification stage there are several shortcomings in the approach that need to be appreciated. The obvious one relates to the level of analysis to be employed. Most human factors practitioners prefer general descriptive levels of analysis rather than fine grain numerical ones (such as the Keystroke level of the GOMS approach). This probably relates to human factors' strengths which lie in qualitative rather than quantitative analysis, and evaluation rather than prediction. This might not be such a problem but for the fact that computer scientists and software engineers strongly distrust human factors' apparently vague suggestions and guidelines. Overcoming this obstacle is not simple and it accounts for many of the difficulties ergonomists have had in influencing designs at the earliest stage.

Less obvious a problem (but one central to the design of electronic documents and hypermedia) however, is that of employing task analysis for situations where innovative products are being developed. By definition, such systems are likely to require skills and procedures for which existing tasks are not directly comparable. In a similar way, building less innovative systems solely on the basis of observed current task performance is likely to limit the potential for improvement and innovation offered by computerizing the task since it locks in old methods to new tools without justification. At a cognitive level, basic representations of the task space need to be maintained to facilitate positive transfer of user knowledge to any new system for similar work. The important point however is to identify the crucial elements to maintain and use them to inform the design of any more innovative aspects. Obviously, detailed knowledge of cognition is a prerequisite of such analysis. Even so, such knowledge would hardly ever be enough.

This point further highlights the need for iteration within the design cycle. As stated earlier, design does not proceed through a finite series of discrete stages but rather requires iteration around the basic triumvirate of specification–design–evaluation. For innovative products suitable task analysis can inform early on but should be backed up with simulations and prototypes for evaluation as soon as possible thereafter. Only in this way can designers be sure of achieving their goal. This point is elaborated further in Shackel (1986) and Diaper (1990). It is unlikely that an ergonomically correct interface will be derived from the first analysis of usability requirements. Rather such work will act to constrain the number of design choices before further analysis and evaluation work is carried out which moves the design closer to its final form.

2.6.4 From analysis to specification and prototype

The leap from information gathered at the earliest stages of design to system specification and subsequent prototyping cannot always be logically determined. For fine grain task analyses, such as the keystroke level analysis of Card *et al.* (1983), the results can logically dictate the choice of design option. However, as most analyses are higher level and some are purely descriptive, such logical decisions concerning usability cannot often be made. However, good task analyses, even in the design of innovative systems can be used to decide between several interface alternatives. For example, knowing that users will be reading large amounts of text on screen could be related to the human factors literature on these issues which offers advice on screen size, resolution, manipulation facilities and so forth thereby constraining the range of suitable alternatives to be considered. From task analysis at least, we would expect a mapping from the described user cognitions and behaviours to specific aspects of the interface.

Stakeholder identifications and user analyses provide less specific routes to interface specification. Generally they inform the process of information gathering at this stage and can be used as supporting evidence to guide decisions taken primarily on the basis of task analysis. Firm boundaries are difficult if not impossible to draw here. The information obtained from any procedure is often similar to that obtained from others, the differences occur in the degree to which the technique concentrates on specific aspects of the users. Ultimately however, any mapping from analysis to interface specification requires some knowledge of human factors.

This non-prescriptive approach to design can be problematic in examining the precise role ergonomic knowledge and methods may have in the design process. Even with the best information available on users, their tasks and the context in which they will perform these tasks, there are still many potential ways in which a designer could proceed. User-centred approaches suggest that the designer should produce a prototype solution and then test this, using the subsequent results to inform the further design of the product until agreed or target levels of usability are achieved.

However, there are potential shortcomings here which need to be addressed. As has been pointed out above, it is unsatisfactory merely to propose prototypes are tested until they get better since theoretically, we have no guarantees they will ever be acceptable. Certainly they should, but the process of improvement is contingent on appropriate testing (outlined above) as a process best left to people with experience in running effective usability trials. Recent evidence from the European software industry (Dillon *et al.*, 1993) suggests that all too often, testing is carried out informally by the designers themselves with minimal or no user involvement. In attempting to change design processes in order to maximize ergonomic input at this stage, the profession is in danger of creating a process with little or none.

Furthermore, unless we have some idea of what the designers are doing with the requirements information in coming up with a prototype, in short understanding design as a social and cognitive process, we are always missing a major variable in the equation (e.g. Curtis, 1990). HCI has attempted to provide designers with guidelines and data in order to influence design for the better but the impact has been minimal. As a field, it can rightly be criticized for its failure to address adequately the nature of design at these levels and it is the author's contention that progress towards truly effective user-centred design processes (and they will be processes since we cannot describe one universal process to cover all artefacts or design tasks) will be hampered until we have a better understanding of these issues.

2.7 Design and science: the product as conjecture

Given a set of requirements and/or information about typical usage, what does a designer or design team do next? This is a deceptively simple looking question to which many would offer the answer: 'they produce a prototype' (or a 'design', a 'model', a 'partial solution', etc.). Yet such an answer tells us little or nothing about the activity of design at this level. Investigations of designers designing are relatively sparse in the literature, although the decisions taken at this stage have an impact out of all proportion with their effort in the product life cycle.

Work with engineers and architects has shed some light on the process. Not surprisingly, engineering design has tackled this issue for many years and there are numerous models for effective design in the literature, as in Archer's (1965) 'linear, critical-path approach' or Asimow's (1962) 'design morphology'. These tended to be prescriptive and sought to guide the designer towards a successful solution rather than reflect the cognitive processes involved. More recently however, the design task (not only or even predominantly software design) has been examined typically as a form of problem solving amenable to analysis by cognitive science.

Part of this literature seems to have uncritically adopted much of the work on cognitive style from mainstream psychology. Cognitive style can loosely be defined as the manner in which people process and respond to information and rests on an assumption that individuals can be distinguished in terms of characteristic processing and response. This has led to the proposal of numerous dimensions of style such as field dependence–independence (Witkin *et al.*, 1977), convergence–divergence (Hudson, 1968) and holism–serialism (Pask, 1976). Few of these putative styles have been shown to predict design performance reliably (or much else for that matter) but the idea of cognitive style remains seductive and has given birth to the notion that designers might manifest distinctive styles of reasoning that could be identified and used to aid design education and to develop computer-aided design (CAD) tools (e.g. Tovey, 1986; Cross, 1985).

Empirical work on design has side-stepped much of this theorizing. Current thought frequently emphasizes the generation–conjecture–analysis model of design proposed by Darke (1979) who interviewed successful architects on their own way of designing and concluded that designers proceed by identifying (often through their own experiences and values as much as by stated requirements) major aims for the design which Darke terms 'the primary generators.' On the basis of these, the designer proposes a partial solution, the 'conjecture', which is subsequently analysed in the light of requirements before the process is repeated. Interestingly, Darke reported that designers in her sample (admittedly only seven) rarely waited for the requirements to be worked out in detail before producing a conjectured solution, most claiming that only with such a solution was it possible to identify many important requirements.

Lawson (1979) compared student scientists with student architects in a problem-solving task involving the arrangement of coloured cubes into a form that would satisfy a stated criterion. He reported that distinctive problem-solving strategies could be identified between these groups. Scientists tended to consider all options before attempting logically to determine an acceptable solution (the so-called logico–deductive approach). Architects adopted a more impulsive strategy, proposing numerous solutions before they completely understood the problem and using the solution as a means to find out more about the target criterion.

Such work is often taken to suggest that designers possess a unique form of cognitive strategy that distinguishes them from scientists but such an argument is less easy to maintain if the context of each group's work is considered more deeply. Designers are judged by their solutions more than their reasoning so it is natural that they seek to demonstrate proposed solutions. Scientists tend to be judged at least in part on their reasoning as much as their answers and the quality of the answer is itself often judged in terms of the reasoning that led to it. Perhaps scientists deal with more abstract entities and merely examining the outputs of scientific work tells us little enough about its intrinsic worth. With designers dealing with physical objects, a proposed solution can quite quickly communicate its own worth directly to an observer. Thus, the strategies manifest in such studies may reflect task variables and experience more than underlying cognitive biases or information processing styles.

What is most interesting is Darke's use of the term 'conjecture' to describe a designer's initial solutions. This invokes Popper's (1972) description of science as the process of conjecture and refutation by which human knowledge accrues. Contrary to the popular image of scientists as rigidly logical, clinically objective, methodical planners (as seems to be suggested by those who (mis)interpret Lawson's oft-cited findings), Popper argues that scientists do, in fact, guess, anticipate and propose tentative solutions which are then subjected to criticism and attempted refutation. According to this philosophy of science, proposed answers and stern tests of their suitability are our only sure way of making progress.

In this way, designers and scientists are less different than many would seek to advocate. What may separate them is less some supposed cognitive characteristic such as style or problem-solving strategy but more the degree to which they seek to rationalize their conjectures in advance and evaluate them once produced. Science has long sought to justify its conjectures in terms of precedent, predicted outcome and theoretical rigour. Similarly, criticism of conjectures acts as a reality check to hold speculative conjectures in perspective. Indeed, from a Popperian perspective, if a theory cannot be put to test such that it can be proven wrong, the theory cannot even be called 'scientific'.

Design practice thus seems similar to science so construed. By extension, a design solution is really just a hypothesis; a conjecture on the part of the designer as to what will suit the intended users or meet the specified requirements. The usability test provides the practical means of refutation. Evidence may suggest that the designers are less rigorous in their test methods than many human factors professionals – here masquerading as the scientists! – would like them to be, but if we accept Popper's model of science, they are both performing very similar activities. Design may be more craft than science but the demarcation between these activities is often subtle and does not deny their similarity. As Popper (1986, p. 87) says:

> there is no clearly marked division between the pre-scientific and the scientific experimental approaches, even though the more and more conscious application of scientific, that is to say critical methods, is of

great importance. Both approaches may be described, fundamentally, as utilizing the method of trial and error. We try, that is, we do not merely register an observation but make active attempts to solve some more of less practical and definite problems. And we make progress if, and only if, we are prepared to learn from our mistakes: to recognize our errors and utilize them critically instead of persevering in them dogmatically. Though this analysis may sound trivial, it describes, I believe, the method of all empirical science. This method assumes a more and more scientific character the more freely and consciously we are prepared to risk a trial and the more critically we watch for the mistakes we always make. And this formula covers not only the method of experiment but also the relationship between theory and experiment. All theories are trials; they are tentative hypotheses, tried out to see if they work; and all experimental corroboration is simply the result of tests undertaken in a critical spirit, in an attempt to find out where our theories err.

Seen in this light, science is very similar to the models of design proposed by Darke (1979), Lawson (1979) and others. Both seek solutions to problems, both conjecture and refute (even if not to similar degrees) and neither can be sure how they ever come up with an idea for a theory, an experiment or a design. If we can conceptualize the design process at the designer level as similar to scientific practice as outlined by Popper, we spend less time searching for uniquely designer modes of thought and address the issue of influencing for the better the conjectures (proposed design solutions) and the systems of refutation (the evaluation methods). In many ways, this is the theme of the present book.

2.8 Electronic documents as a technology?

At the start of this chapter it was stated that electronic texts were just one more application of computers and a relatively natural one at that. Talk of documents as a technology might be disconcerting at first but this probably results more from an old-fashioned view of what a technology is rather than any flaw in the description. Certainly if we can talk of information technology as more than the technical or applied scientific aspects of information handling, in fact as being the physical instantiation of these principles, then a document is a technology for reading and electronic texts are just another example of reading technology.

Who then are the stakeholders and users of this technology? Almost everyone reads and ideally everyone should be able to do so. Parents, educators, governments and certain employers expend vast amounts of capital and effort in ensuring basic levels of literacy are maintained and hopefully maximized. To the extent that almost everyone is a reader, then almost everyone is a user or potential user of this technology – and to the extent that society places great store by literacy, everyone is a stakeholder in information technologies.

The very breadth of the user population can mislead some people into thinking that we need not analyse the user population for this technology much further since the users are 'everyone and anyone'. This is a mistake. The diversity of the user population in total means that they embody almost all sources of individual difference imaginable in the human species: language ability, intelligence, experience, task knowledge, age, education, cognitive style, etc. As stated in relation to contextual issues, the specific nature of the user population, to which any one document is aimed, needs to be made explicit in any discussion of its usability.

We would expect to find users of documents in work situations, school and universities, homes, public environments and so forth. The type of document will range from technical to the leisure material, books and magazines, articles and essays, notes and memos, bills and love-letters, etc. Users will be reading for pleasure and for need, to learn and to amuse, to find specific information and to browse, to pass exams and to pass the time. Thus, when talking about electronic documents and their usability it is essential that the contextual variables are made explicit and we avoid the trap of endorsing or dismissing the medium on the basis of an excellent or an inappropriate design for one specific context. These issues will become clearer as the book progresses and specific document design issues are examined.

Notes

1. In the USA this acronym was often shortened to CHI (computer–human interaction), a reversal that some human scientists are quick to point to as an example of technologists putting the computer before the human!
2. The suffix 'friendly' has been attached to so many nouns that it is easy to forget that it came to prominence in recent times as an ergonomic decriptor of software. I have seen various processes, objects and events described in the last few years as 'eater-friendly' (a variety of bean!), 'customer-friendly' (a shop), 'audience-friendly' (a theatre) and 'cyclist-friendly' (a road!) that it is clear that the philosophy of user-centredness has at least been hijacked by marketing professionals.
3. One could add a host of other reasons why user-centredness should not be equated merely with rapid prototyping such as the changing form of user requirements over time, the emergence of other requirements only when a system has been used for real, the absence of functionality in prototypes, the validity of testing prototypes and the tendency for designers to preserve rather than reject prototyped interfaces.
4. An example might be a distinction between users in terms of 'office' or 'shop-floor' worker. These descriptors appear to distinguish users on little other than a label but actually inform us of potential environmental constraints that one should examine closer.

3

So what do we know?
An overview of the empirical literature on
reading from screens

Books are a load of crap.

Philip Larkin, *Study of Reading Habits* (1964)

3.1 Introduction

In simple terms, there exist two schools of thought on the subject of electronic
texts. The first holds that paper is far superior and will never be replaced by
screens. The argument is frequently supported by reference either to the type
of reading scenarios that would currently prove difficult if not impossible to
support acceptably with electronic text – reading a newspaper on the beach or
browsing a magazine in bed – or the unique tactile qualities of paper. The
latter aspect is summed up neatly in Garland's (1982) comment that electronic
text may have potential uses: 'but a book is a book is a *book*. A reassuring,
feel-the-weight, take-your-own-time kind of thing . . .' (cited in Whaller, 1987,
p. 261).

The second school favours the use of electronic text, citing ease of storage
and retrieval, flexibility of structure and saving of natural resources as major
incentives. According to this perspective, electronic text will soon replace
paper and in a short time (usually ten years hence) we shall all be reading from
screens as a matter of habit. In the words of its greatest proponent, Ted Nelson
(1987), 'the question is not can we do everything on screens, but when will we,
how will we and how can we make it great? This is an article of faith – its
simple obviousness defies argument. If you don't get it there is no persuading
you; if you do get it you don't need to be persuaded.'[1]

Such extremist positions show no signs of abating though it is becoming
clear to many researchers in the domain that neither is particularly
satisfactory. Reading from screens *is* different from paper and there are many
scenarios such as those cited that current technology would not support well, if
at all. However, technology is developing and electronic text of the future is

unlikely to be handicapped by limitations in screen image and portability that currently seem major obstacles. As Licklider pointed out when considering the application of computers in libraries as early as 1965, 'our thinking and our planning need not be, and indeed should not be, limited by literal interpretation of the existing technology' (p. 19).

Even so, paper is an information carrier *par excellence* and possesses an intimacy of interaction that can never be obtained in a medium that by definition imposes a microchip interface between the reader and the text. Furthermore, the millions of books that exist now will not all find their way into electronic form, thus ensuring the existence of paper documentation for many years yet.

The aim of the present chapter is not to resolve the issue of whether one or other medium will dominate but to examine critically the reported differences between them in terms of use and thereby support reasoned analysis of the paper versus electronic text debate from the perspective of the reader. The intention is not to support prejudices or to allow people to 'get' the message (in Nelson-speak) but to let the evidence do the talking. In so doing, it indicates the approach taken to date by most investigators of electronic text and thus where such work seeks to lead document designers.

3.2 The outline of the review

At the outset it must be stated that drawing any firm conclusions from the literature is difficult. Helander *et al.* (1984) evaluated 82 studies concerning human factors research on VDUs and concluded (p. 55):

> Lack of scientific rigour has reduced the value of many of these studies. Especially frequent were flaws in experimental design and subject selection, both of which threaten the validity of results. In addition, the choice of experimental settings and dependent and independent variables often made it difficult to generalize the results beyond the conditions of the particular study.

This is not an isolated criticism of the literature Waern and Rollenhagen (1983) point to the frequently narrow scope of experimental designs in such studies. Important factors are either not properly controlled or are simply not reported and most studies use unique procedures and equipment, rendering direct comparison meaningless. Similarly, Dillon *et al.* (1988) drew attention to the limited experimental tasks employed by many investigators. Despite these problems, a review of experimental evidence is the most appropriate place to start any work of this kind.

A detailed literature already exists on typographical issues related to text presentation on paper (see particularly the work of Tinker, 1958, 1963) and issues such as line spacing and formatting are well researched. This work will not be reviewed here as much of it remains unreplicated on VDUs and evidence to date suggests that, even when such factors are held constant, reading differences between the two presentation media remain (see for example Creed *et al.*, 1987).

In the first instance this review examines the nature of the possible differences between the media and draws a distinction here between outcome (section 3.4) and process (section 3.5) differences. Following this, a brief overview of the type of research that has been carried out is presented (section 3.6). Experimental comparisons of reading from paper and screen are then reviewed. A final section highlights the shortcomings of much of this work and indicates the way forward for research in this domain.

3.3 Observed differences: outcome and process measures

Analysing a complex human activity such as reading is not simple and a distinction has been drawn between assessing reading behaviour in terms of outcome and process measures (Shumacher and Waller, 1985). Outcome measures concentrate on the results of reading and considers such variables as: amount of information retrieved, accuracy of recall, time taken to read the text, spelling and syntactical errors identified and so forth. Process measures are more concerned with what is happening when readers use a text and include such variables as where the readers look in the text, how they manipulate it and how knowledge of contents is gained.

In the domain of electronic text, outcome measures take on particular relevance as advocates continually proclaim increased efficiency and improved performance (i.e. reading outcomes) with computer presented material (aspects of obvious concern to ergonomists). It is not surprising therefore to find that most work comparing the two media has concentrated heavily on such differences. With the emergence of hypertext however, navigation has become a major issue and process measures are gaining increased recognition of importance.

In the following sections, a summary of the observed differences between the media in terms of outcomes and processes is presented. This is not an exhaustive review however as the field is wide and new data are being reported with almost every issue of a relevant journal or proceedings of relevant conferences. It is intended more as a perspective on this literature and a way of categorizing the emerging findings. As such, it relies heavily on earlier reviews of the literature by the author and colleagues (Dillon, *et al.*, 1988; Dillon, 1992).

3.4 Outcome measures

3.4.1 Speed

By far the most common experimental finding used to be that silent reading from screen is significantly slower than reading from paper (Kak, 1981; Muter *et al.*, 1982; Wright and Lickorish, 1983; Gould and Grischkowsky, 1984; Smedshammar *et al.*, 1989; Smith and Savory, 1989). Figures vary according to means of calculation and experimental design but the evidence suggests a performance deficit of 20–30 per cent when reading from screen.

However, despite the apparent similarity of findings, it is not clear whether the same mechanisms have been responsible for the slower speed in these experiments, given the great disparity in procedures. For example, in the study by Muter *et al.* (1982), subjects read white text on a blue background, with the subject being approximately 5 m from the screen. The characters, displayed in teletext format on a television, were approximately 1 cm high, and time to fill the screen was approximately 9 s. In comparison, Gould and Grischkowsky (1984) used greenish text on a dark background. Characters were 3 mm high and subjects could sit at any distance from the screen. They were encouraged to adjust the room lighting level and the luminance and contrast of the screen for their comfort. Printed text used 4 mm characters and was laid out identically to the screen text. Wright and Lickorish (1983) displayed text as white characters on a black 12inch screen driven by an Apple II microcomputer with lower case facility. This would suggest that it was closer to Gould's text than Muter's text in appearance. Printed texts were photocopies of printouts of the screen displays produced on an Epson MX-80 dot matrix printer, compared with Gould's 10-point monospace Letter Gothic font.

In contrast to these studies, Muter and Maurutto (1991), Askwall (1985) and Cushman (1986) found that reading speed was unaffected by the presentation medium. Askwall attributes this difference in findings to the fact that her texts were comparatively short (22 sentences), and the general lack of experimental detail makes alternative interpretations difficult. Although it is reported that a screen size of 24 rows by 40 columns was used, with letter size approximately 0.5×0.5 cm and viewing distance of approximately 30–50 cm, no details of screen colour or image polarity and none of the physical attributes of the printed text are given.

Cushman measured reading speed and comprehension during 80-minute reading sessions. Negative and positive image VDU and microfiche presentations were used and most of the 76 subjects are described as having had 'some previous experience using microfilm readers and VDUs'. On the basis of this study Cushman concluded that there was no evidence of a performance deficit for the VDU presentations compared with printed paper.

Leventhal *et al.* (1993) examined readers locating information in an encyclopedia presented in hypertext (run on an Apple Macintosh but with no details of screen provided) or paper form. They report a 'marginally significant' effect for speed favouring paper (the figures published indicate that paper readers were on average about 15 per cent faster than screen readers) but the authors stress that the nature of the information location task was crucial.

As this indicates, the evidence surrounding the argument for a speed deficit in reading from VDUs is less than conclusive. A number of intervening variables, such as the size, type and quality of the VDU may have contaminated the results and complicate their interpretation. Gould *et al.* (1987a) noted, many of these experiments are open to interpretation but 'the evidence on balance . . . indicates that the basic finding is robust – people do read more slowly from CRT displays' (p. 269).

However, as typical displays evolve in quality terms this may not hold true indefinitely but to date, there is probably insufficient evidence around to overturn the summary of Gould *et al.*

3.4.2 Accuracy

Accuracy of reading could refer to any number of everyday activities such as locating information in a text, recalling the content of certain sections and so forth. In experimental investigations of reading from screens the term accuracy has several meanings too though it most commonly refers to an individual's ability to identify errors in a proof-reading exercise. While studies have been carried out which failed to report accuracy differences between VDUs and paper (e.g. Wright and Lickorish, 1983; Gould and Grischkowsky, 1984) well controlled experiments by Creed *et al.* (1987) and Wilkinson and Robinshaw (1987) report significantly poorer accuracy for such proof-reading tasks on screens.

Since evidence for the effects of presentation media on such accuracy measures often emerges from the same investigations which looked at the speed question, the criticisms of procedure and methodology outlined above apply equally here. The measures of accuracy employed also vary. Gould and Grischkowsky (1984) for example required subjects to identify misspellings of four types: letter omissions, substitutions, transpositions and additions, randomly inserted at a rate of one per 150 words. Wilkinson and Robinshaw (1987) argue that such a task hardly equates to true proof-reading but is merely identification of spelling mistakes. In their study they tried to avoid spelling or contextual mistakes and used errors of five types: missing or additional spaces, missing or additional letters, double or triple reversions, misfits or inappropriate characters, and missing or inappropriate capitals. It is not always clear why some of these error types are not spelling or contextual mistakes but Wilkinson and Robinshaw suggest their approach is more relevant to the task demands of proof-reading than Gould and Grischkowsky's.

However Creed *et al.* (1987) distinguished between visually similar errors (e.g. 'e' replaced by 'c'), visually dissimilar errors (e.g. 'e' replaced by 'w') and syntactic errors (e.g. 'gave' replaced by 'given'). They argue that visually similar and dissimilar errors require visual discrimination for identification while syntactic errors rely on knowledge of the grammatical correctness of the passage for detection and are therefore more cognitively demanding. This error classification was developed in response to what they saw as the shortcomings of the more typical accuracy measures which provide only gross information concerning the factors affecting accurate performance. Their findings indicate that visually dissimilar errors are significantly easier to locate than either visually similar or syntactic errors.

In a widely reported study Egan *et al.* (1989) compared students' performance on a set of tasks involving a statistics text presented on paper or screen. Students used either the standard textbook or a hypertext version run on SuperBook, a structured browsing system, to search for specific information in the text and write essays with the text open. Incidental learning and subjective ratings were also assessed. The search tasks provide an alternative to, and more realistic measure of reading accuracy than identifying spelling errors.

The authors report that subjects using the hypertext performed significantly more accurately than those using the paper text. However a closer look at the experiment is revealing. With respect to the search tasks, the questions posed were varied so that their wording mentioned terms contained in the body of the text, in the headings, in both of these or neither. Not surprisingly the largest advantage to electronic text was observed where the target information was only mentioned in the body of text (i.e. there were no headings referring to it). Here the search facility of the computer outperformed humans. When the task was less biased against the paper condition e.g. searching for information to which there are headings, no significant difference was observed. Interestingly the poorest performance of all was for SuperBook users searching for information when the question did not contain specific references to words used anywhere in the text. In the absence of suitable search parameters or look-up terms hypertext suddenly seemed less effective.

McKnight *et al.* (1990a) compared reading in two versions of hypertext, a word processor file and a paper copy of a document on winemaking. The measure of accuracy taken was the number of answers correctly made to a set of questions seeking information to be found in the document. Interestingly they report no significant difference between paper and word processor file, but readers in both hypertext conditions were significantly less accurate than readers of the paper document. In a subsequent study, McKnight *et al.* (1992) report that readers of a hypertext document complained more than readers of paper about a lack of confidence in finding all relevant material.

Leventhal *et al.* (1993) performed a detailed study of question–answering with an encyclopedia presented as hypertext or paper and carefully manipulated the type of questions asked in order to ensure a reliable test of distinct information location strategies. Their questions examined factual information located in titles, in the body of text sections, or in maps as well as questions requiring inference from material in the body of the text or in the maps. Interestingly, they report an effect for question type which interacted with presentation medium but no significant effect for medium alone. Readers of the paper version had most difficulties locating information in the body of the text that had no key to location in the question (as in the Egan *et al.* study) but performed best on questions requiring inference and location in maps. All subjects also displayed an improvement with practice that was independent of presentation medium.

Regardless of the interpretation that is put on the results of any one of these studies, the fact remains that investigations of reading accuracy from VDU and paper take a variety of measures as indications of performance. Therefore two studies, both purporting to investigate reading accuracy may not necessarily measure the same events. The issue of accuracy is further complicated by the presence or absence of certain enabling facilities (e.g. search routines) and the potential to alter information structures with hypertext applications. Obviously, for some tasks, the computer is the only viable presentation medium, but the more the task involves serial reading of text and less manipulation and searching, paper retains advantages related to letter discrimination. Before claims are made about the relative accuracy of either medium, clear task descriptions need to be provided.

3.4.3 Fatigue

The proliferation of information technology has traditionally brought with it fears of harmful or negative side-effects for users who spend a lot of time in front of a VDU (e.g. Pearce, 1984). In the area of screen reading this has manifested itself in speculation of increased visual fatigue and/or eyestrain when reading from screens as opposed to paper.

In the Muter *et al.* (1982) study subjects were requested to complete a rating scale on a number of measures of discomfort including fatigue and eyestrain both before and after exposure to the task. There were no significant differences reported on any of these scales either as a result of condition or time. Similarly Gould and Grischkowsky (1984) obtained responses to a 16-item *Feelings Questionnaire* after each of six 45-minute work periods. This questionnaire required subjects to rate their fatigue, levels of tension, mental stress and so forth. Furthermore various visual measurements such as flicker and contrast sensitivity, visual acuity and phoria, were taken at the beginning of the day and after each work period. Neither questionnaire responses nor visual measures showed a significant effect for presentation medium. These results led the authors to conclude that good-quality VDUs in themselves do not produce fatiguing effects, citing Starr (1984) and Sauter *et al.* (1983) as supporting evidence.

In a more specific investigation of fatigue Cushman (1986) investigated reading from microfiche as well as paper and VDUs with positive and negative image. He distinguished between visual and general fatigue, assessing the former with the visual fatigue graphic rating scale (VFGRS) which subjects use to rate their ocular discomfort, and the latter with the feeling-tone checklist (FTC) (Pearson and Byars, 1956). With respect to the VDU conditions, the VFGRS was administered before the session and after 15, 30, 45 and 60 minutes as well as at the end of the trial at 80 minutes. The FTC was completed before and after the session. The results indicated that reading from positive presentation VDUs (dark characters on light background) was more fatiguing than paper and leads to greater ocular discomfort than reading from negative presentation VDUs.

Cushman explained the apparent conflict of these results with the established literature in terms of the refresh rate of the VDUs employed (60 Hz) which may not have been enough to completely eliminate flicker in the case of positive presentation, a suspected cause of visual fatigue. Wilkinson and Robinshaw (1987) also reported significantly higher fatigue for VDU reading and while their equipment may also have influenced the finding, they dismiss this as a reasonable explanation on the grounds that no subject reported lack of clarity or flicker and their monitor was typical of the type of VDU that users find themselves reading from. They suggest that Gould and Grischkowsky's (1984) equipment was 'too good to show any disadvantage' and that their method of measuring fatigue was artificial. By gathering information after a task and across a working day Gould and Grischkowsky missed the effects of fatigue within a task session and allowed time of day effects to contaminate the results. Wilkinson and Robinshaw liken the proof-reading task used in these studies to vigilance performance and argued that fatigue is more likely to occur within the single work period where there are no rest pauses allowing recovery. Their results showed a performance decrement across the 50-minute task employed, leading them to conclude that reading from typical VDUs at least for periods longer than 10 minutes is likely to lead to greater fatigue.

It is not clear how comparable conclusions drawn from measures of fatigue such as subjective ratings of ocular discomfort are with inferences drawn from performance rates. It would seem safe to conclude that users do not find reading from VDUs intrinsically fatiguing but that performance levels may be more difficult to sustain over time when reading from average quality screens. As screen standards increase over time this problem should be minimized.

3.4.4 Comprehension

Perhaps more important than the questions of speed and accuracy of reading is the effect of presentation medium on comprehension. As Dillon *et al.* (1988) indicated, should any causal relationship ever be identified between reading from VDU and reduced comprehension, the impact of this technology would be severely limited. The issue of comprehension has not been as fully researched as one might expect, perhaps in no small way due to the difficulty of devising a suitable means of quantification, i.e. how does one measure a reader's comprehension?

Post-task questions about content of the reading material are perhaps the simplest method of assessment, although care must be taken to ensure that the questions do not simply demand recall skills. Muter *et al.* (1982) required subjects to answer 25 multiple-choice questions after two 1-hour reading sessions. Due to variations in the amount of material read by all subjects, analysis was reduced to responses to the first eight questions of each set. No effect on comprehension was found either for condition or question set. Kak (1981) presented subjects with a standardized reading test (the Nelson–Denny test) on paper and VDU. Comprehension questions were answered by hand. No significant effect for presentation medium was observed. A similar result was found by Cushman (1986) in his comparison of paper, microfiche and VDUs. Interestingly however, he noted a negative correlation between reading speed and comprehension, i.e. comprehension tended to be higher for slower readers.

Belmore (1985) asked subjects to read short passages from screen and paper and measured reading time and comprehension. An initial examination of the results appeared to show a considerable disadvantage, in terms of both comprehension and speed, for screen presented text. However, further analysis showed that the effect was only found when subjects experienced the screen condition first. Belmore suggested that the performance decrement was due to the subjects' lack of familiarity with computers and reading from screens – a factor commonly found in this type of study. Very few of the studies reported here attempted to use a sample of regular computer users.

Gould *et al.* (1987a) compared subjects reading for comprehension with proof-reading for both media in order to check that typical proof-reading tasks did not intrinsically favour a medium that supported better character discrimination. Though only concerned with reading speed (i.e. they took no comprehension measures) they found that reading for comprehension actually exacerbated the differences between paper and screen.

The Egan *et al.* study (1989) described earlier required subjects to write essay type answers to open book questions using paper or hypertext versions of a statistics book. Experts rated the essays and it was observed that users of the hypertext version scored significantly higher marks than users of the paper book. Thus, the authors conclude, the potential of restructuring the text with current technology can significantly improve comprehension for certain tasks.

A study covering this issue by Muter and Maurutto (1991) asked readers to answer questions about a short story read either on paper or screen immediately after finishing the reading task. They reported no significant comprehension difference between readers using either medium. Similarly, McKnight *et al.* (1992) compared comprehension as assessed by essays written on the basis of notes made during exposure to material presented via hypertext or book. Readers were students taking a postgraduate course of study in this area and the essays were marked by a domain expert. Again, they reported no significant comprehension difference between the media. However, Leventhal *et al.* (1993) tested incidental learning after their study by asking subjects to identify titles of sections they had read. They found significantly more incidental learning in the hypertext condition than paper.

It seems therefore that comprehension of material is not negatively affected by presentation medium and under some circumstances may even be improved. However, a strong qualification of this interpretation of the experimental findings is that suitable comprehension measures for reading material are difficult to devise. The expert rating used by Egan *et al.* or McKnight *et al.* (1992) is ecologically valid in that it conforms to the type of assessment usually employed in schools and colleges but the sensitivity of post-task question and answer sessions or essay writing to subtle cognitive differences caused by presentation medium is debatable. Without evidence to the contrary though, it would seem as if reading from VDUs does not negatively affect comprehension rates though it may affect the speed with which readers can attain a given level of comprehension.

3.4.5 Preference

Part of the folklore of human factors research is that naïve users tend to dislike using computers and much research aims at encouraging user acceptance of systems through more usable interface design. Given that much of the evidence cited here is based on studies of relatively novice users it is possible that the results are contaminated by subjects' negative predispositions towards reading from screen. On the basis of a study of 800 VDU operators' comparisons of the relative qualities of paper and screen-based text, Cakir *et al.* (1980) report that high quality typewritten hardcopy is generally judged to be superior. Preference ratings were also recorded in the Muter *et al.* (1982) study and despite the rather artificial screen reading situation tested, users only expressed a mild preference for reading from a book. They expressed the main advantage of book reading to be the ability to turn back pages and re-read previously read material, mistakenly assuming that the screen condition prevented this.

Starr (1984) concluded that relative subjective evaluations of VDUs and paper are highly dependent on the quality of the paper document, though one may add that the quality of the VDU display probably has something to do with it too. Egan *et al.* (1989) found a preference for hypertext over paper amongst subjects in their study of a statistics text where the electronic copy was displayed on a very high quality screen. Recent evidence from Muter and Maurutto (1991) revealed that approximately 50 per cent of subjects in their comparative studies of reading from paper and current screens expressed a preference for screen; a similar figure was reported by Spinks (1991) but Leventhal *et al.* (1993) report that subjective ratings of task enjoyment were significantly higher for hypertext readers in their comparative study. Such findings lend some support to the argument that preferences are shifting as screen technology improves or as readers become more familiar with the technology.

What seems to have been overlooked as far as formal investigation is concerned is the natural flexibility of books and paper over VDUs, e.g. paper documents are portable, cheap, apparently 'natural' in our culture, personal and easy to use. The extent to which such common-sense variables influence user performance and preferences is not yet well-understood.

3.4.6 Summary

Empirical investigations of the area have suggested five possible outcome differences between reading from screens and paper. As a result of the variety of methodologies, procedures and stimulus materials employed in these studies, definitive conclusions cannot be drawn. It seems certain that reading speeds are reduced on typical VDUs and accuracy may be lessened for visually and cognitively demanding tasks. Fears of increased visual fatigue and reduced levels of comprehension as a result of reading from VDUs have gained little empirical support though the validity of separating accuracy and comprehension into two discrete outcomes is debatable. With respect to reader preference, top quality hardcopy seems to be preferred to screen displays, which is not altogether surprising but shifts may be occurring as technology improves and readers become more experienced with the new medium.

3.5 Process measures

Without doubt, the main obstacle to obtaining accurate process data is devising a suitable, non-intrusive observation method. While techniques for measuring eye-movements during reading now exist, it is not at all clear from eye-movement records what the reader was thinking or trying to do at any time. Furthermore, use of such equipment is rarely non-intrusive, often requiring the reader to remain immobile through the use of head restraints, bite bars etc., or read the text one line at a time from a computer display – hardly equatable to normal reading conditions!

Less intrusive methods such as the use of light pens in darkened environments to highlight the portion of the text currently viewed (Whalley and Fleming, 1975) or modified reading stands with semi-silvered glass which reflect the readers' eye-movements in terms of current text position to a video camera (Pugh, 1979) are examples of the lengths researchers have gone to in recording the reading process. However, none of these are ideal as they alter the reading environment, sometimes drastically, and only their staunchest advocate would describe them as non-intrusive.

Verbal protocols of people interacting with texts require no elaborate equipment and can be elicited wherever a subject normally reads. Thus the technique is cheap, relatively naturalistic and physically non-intrusive. However, verbal protocol analysis has been criticised for interfering with the normal cognitive processing involved in task performance and requiring the presence of an experimenter to sustain and record the verbal protocol (Nisbett and Wilson, 1977).

Although a perfect method does not yet exist it is important to understand the relative merits of those that are available. Eye-movement records have significantly aided theoretical developments in modelling reading (e.g. Just and Carpenter, 1980) while use of the light-pen-type techniques has demonstrated their worth in identifying the effects of various typographic cues on reading behaviour (e.g. Waller, 1984). Verbal protocols have been effectively used by researchers to gain information on reading strategies (e.g. Olshavsky, 1977).

Nevertheless, such techniques have rarely been employed with the intention of assessing the process differences between reading from paper and from screen. Where paper and hypertext are directly compared, although process measures may be taken with the computer and or video cameras, the final comparison often rests on outcome measures (e.g. McKnight *et al.*, 1990a).

Despite this, it is widely accepted that the reading process with screens is different than that with paper regardless of any outcome differences. The following sections outline three of the most commonly cited process differences between the media. In contrast to the outcome differences it will be noted that, for the reasons outlined above, these differences are less clearly empirically demonstrated.

3.5.1 Eye movements

Mills and Weldon (1986) argue that measures of eye movements reflect difficulty, discriminability and comprehensibility of text and can therefore be used as a method of assessing the cognitive effort involved in reading text from paper or screen. Indeed Tinker (1958) reports how certain text characteristics affect eye movements and Kolers *et al.* (1981) employed measures of eye movement to investigate the effect of text density on ocular work and reading efficiency. Obviously if reading from screen is different than paper then noticeable effects in eye-movement patterns might be found indicating possible causes and means of improvement.

Eye movements during reading are characterized by a series of jumps and fixations. The latter are of approximately 250 msec duration and it is during these that word perception occurs. The 'visual reading field' is the term used to describe that portion of foveal and parafoveal vision from which visual information can be extracted during a fixation and in the context of reading, this can be expressed in terms of the number of characters available during a fixation. The visual reading field is subject to interference from text on adjacent lines, the effect of which seems to be a reduction in the number of characters available in any given fixation and hence a reduction in reading speed.

Gould *et al.* (1987a) report an investigation of eye-movement patterns when reading from either medium. Using a photoelectric eye-movement monitoring system, subjects were required to read two 10-page articles, one on paper, the other on screen. Eye movements typically consisted of a series of fixations on a line, with re-fixations and skipped lines being rare. Movement patterns were classified into four types: fixations, undershoots, regressions and re-fixations. Analysis revealed that when reading from VDU subjects made significantly more (15 per cent) forward fixations per line. However this 15 per cent difference translated into only 1 fixation per line. Generally, eye-movement patterns were similar and no difference in duration was observed. Gould *et al.* explained the 15 per cent fixation difference in terms of image quality variables. Interestingly they report that there was no evidence that subjects lost their place, 'turned-off' or re-fixated more when reading from VDUs.

It seems therefore that gross differences in eye movements do not occur between screen and paper reading. However, given the known effect of typographic cueing on eye movements with paper and the oft-stated non-transferability of paper design guidelines to screens, it is possible that hypertext formats might influence the reading process at this level in a manner worthy of investigation.

3.5.2 Manipulation

Perhaps the most obvious difference between reading from paper and from screens is the ease with which paper can be manipulated and the corresponding difficulty of so doing with electronic text. Yet manipulation is an intrinsic part of the reading process for most tasks. Manipulating paper is achieved by manual dexterity, using fingers to turn pages, keeping one finger in a section as a location aid, or flicking through tens of pages while browsing the contents of a document, activities difficult or impossible to support electronically (Kerr, 1986).

Such skills are acquired early in a reader's life and the standard physical format of most documents means these skills are transferable between all document types. With electronic text this does not hold. Lack of standards means that there is a bewildering range of interfaces to computer systems and mastery of manipulation in one application is no guarantee of an ability to use another. Progressing through the electronic document might involve using a mouse and scroll bar in one application and function keys in another; one might require menu selection and 'page' numbers while another supports touch-sensitive 'buttons'. With hypertext, manipulation of large electronic texts can be rapid and simple while other systems might take several seconds to refresh the screen after the execution of a 'next page' command.

Such differences will almost certainly affect the reading process. Waller (1986) suggests that as readers need to articulate their needs in manipulating electronic texts (i.e. formulate an input to the computer to move the text rather than directly and automatically performing the action themselves) a drain on the cognitive resources required for comprehension could occur. Richardson *et al.* (1988) report that subjects find text manipulation on screen awkward compared to paper, stating that the replacement of direct manual interaction with an input device deprived users of much feedback and control.

It is obvious that manipulation differences exist and that electronic text is usually seen as the less manipulable medium. Current hypertext applications however, support rapid movement between various sections of text which suggests that innovative manipulations might emerge that, once familiar with them, convey advantages to the reader of electronic texts. This is an area for further work.

3.5.3 Navigation

When reading a lengthy document the reader will need to find their way through the information in a manner that has been likened to navigating a physical environment (Dillon *et al.*, 1993). There is a striking consensus among many researchers in the field that this process is the single greatest difficulty for readers of electronic text. This is particularly (but not uniquely) the case with hypertext where frequent reference is made to 'getting lost in hyperspace' (e.g. Conklin, 1987; McAleese, 1989) which is described as in the oft-quoted line 'the user not having a clear conception of the relationships within the system or knowing his present location in the system relative to the display structure and finding it difficult to decide where to look next within the system' (Elm and Woods, 1985, p. 927).

With paper documents there tends to be at least some standards in terms of organization. With books for example, contents pages are usually at the front, indices at the back and both offer some information on where items are located in the body of the text. Concepts of relative position in the text such as 'before' and 'after' have tangible physical correlates. No such correlation holds with hypertext and such concepts are greatly diminished in standard electronic text.

There is some direct empirical evidence in the literature to support the view that navigation can be a problem. Edwards and Hardman (1989) for example, describe a study which required subjects to search through a specially designed hypertext. In total, half the subjects reported feeling lost at some stage (this proportion is inferred from the data reported). Such feelings were mainly due to 'not knowing where to go next' or 'not knowing where they were in relation to the overall structure of the document' rather than 'knowing where to go but not knowing how to get there' (descriptors provided by the authors). Unfortunately, without direct comparison of ratings from subjects reading a paper equivalent we cannot be sure such proportions are solely due to using hypertext.

McKnight *et al.* (1990a) compared navigation for paper, word processor and two hypertext documents by examining the number of times readers went to index and contents pages/sections, inferring that time spent here gave an indication of navigation problems. They reported significant differences between paper and both hypertext conditions (the latter proving worse), with word processor users spending about twice as long as paper readers in these sections.

Indirect evidence comes from the numerous studies which have indicated that users have difficulties with a hypertext (Monk *et al.*, 1988; Gordon *et al.*, 1988). Hammond and Allinson (1989, p. 294) speak for many when they say: 'experience with using hypertext systems has revealed a number of problems for users . . . . First, users get lost . . . . Second, users may find it difficult to gain an overview of the material . . . . Third, even if users know specific information is present they may have difficulty finding it.'

There are a few dissenting voices. Brown (1988, p. 2) argues that 'although getting lost is often claimed to be a great problem, the evidence is largely circumstantial and conflicting. In some smallish applications it is not a major problem at all.'

This quote is telling in several ways. The evidence for navigational difficulties *is* often circumstantial, as noted above. The applications in which Brown claims it is not a problem at all, are, to use his word, 'smallish' and this raises a crucial issue with respect to electronic text research that is taken up later, how much faith can we place in evidence from studies involving very short texts. However, the evidence that we currently possess seems to indicate that navigation in the electronic medium is a reading process issue worthy of further investigation.

3.5.4 Summary

The reading process is affected by the medium of presentation though it is extremely difficult to quantify and demonstrate such differences empirically. The major differences appear to occur in manipulation which seems more awkward with electronic texts and navigation which seems to be more difficult with electronic and particularly hypertexts. Eye-movement patterns do not seem to be significantly altered by presentation medium. Further process issues may emerge as our knowledge and conceptualization of the reading process improves.

3.6 Explaining the differences: a classification of issues

While the precise nature and extent of the differences between reading from either medium have not been completely defined, attempts to identify possible causes of any difference have frequently been made. A significant literature exists on issues dealing with display characteristics such as line length and spacing. It is not the aim of this review to detail this literature fully except where it relates to possible causes for reading differences between paper and screen. Experimental investigations which have controlled such variables have still found performance deficits on VDUs, thus suggesting that the root cause of observed differences lies elsewhere. For a comprehensive review of these issues see Mills and Weldon (1986).

From an ergonomic perspective the human is generally considered to respond to stimulation at three levels, physically, perceptually and cognitively. It is not surprising therefore that in seeking to understand better the nature of the differences between the media that researchers have sought explanations at these levels.

In the case of reading it is clear that documents must be handled and manipulated (physical level) before the reader can visually perceive the written information (perceptual level) and thus make sense of the material it contains (cognitive level). In a very real sense all these areas are inter-dependent and the boundaries therefore between the levels is fuzzy rather than rigid. However, 'divide and conquer' is a principle that has long served scientists and it is to be expected that in seeking to explain media induced effects in the reading process, researchers have sought to tackle the problem this way.

A complete trawl through the experimental literature at these levels would be both exhausting to the reader (not to mention the author) and not best serve our interests here. Furthermore, detailed reviews of large sections of this literature have already been produced (e.g. Dillon, 1992). In the present chapter therefore a brief summary of the findings of experimental literature on the subject will be outlined in a form that loosely adopts the three-level categorization outlined above.

The emphasis on empirical literature is important. There is no end to the authors who seek to pass opinions or subjective interpretations of the effect of screen presentation on reading. For the most part their opinions are based on very little except their own experiences and if the history of user-centred design has shown nothing else, it has at least demonstrated that the personal experience or opinion of any one advocate is usually of little or no use in explaining complex issues of user psychology.

3.7 Physical sources of difference

An electronic text is physically different from a paper one. Consequently, many researchers have examined these aspects of the medium in an attempt to explain the performance differences. On an obvious level books look different, they afford clues to age, size and subject matter from just the simplest of examinations. They are also easily portable. However, apart from these factors, there are potentially more subtle ergonomic issues involved in reading from screens which some researchers have sought to examine experimentally.

3.7.1 Orientation

One of the advantages of paper over VDUs is that it can be picked up and oriented to suit the reader. VDUs present the reader with only limited flexibility to alter vertical or horizontal orientation. Gould *et al.* (1987a) investigated the hypothesis that differences in orientation may account for differences in reading performance. Subjects were required to read three articles, one on a vertically positioned VDU, one on paper-horizontal and the other on paper-vertical (paper attached via copy-holder to equivalent VDU). Both paper conditions were read significantly faster than the VDU and there were no accuracy differences. While orientation has been shown to affect reading rate of printed material (Tinker, 1963) it does not seem to account for the observed reading differences in the comparisons reported here.

3.7.2 Aspect ratio

The term 'aspect ratio' refers to the relationship of width to height. Typical paper sizes are higher than they are wide, while the opposite is true for typical VDU displays. Changing the aspect ratio of a visual field may affect eye-movement patterns sufficiently to account for some of the performance differences. Gould (1986) had 18 subjects read three 8-page articles on VDU, paper and paper-rotated (aspect ratio altered to resemble screen presentation). The results however showed little effect for ratio.

3.7.3 Handling and manipulation

Not only are books relatively easy to handle and manipulate, but one set of handling and manipulation skills serves to operate most document types. Electronic documents however cannot be handled directly at all but must be presented through a computer screen and even then, the operations one can perform and the manner in which they are performed can vary tremendously from presentation system to presentation system.

Over the last 15 years numerous input devices have been designed and proposed as optimal for computer users, e.g. mouse, function keyboard, joystick, light pen, etc. Since the claim of Card *et al*. (1978) that the speed of text selection via a mouse was constrained only by the limits of human information processing, this device has assumed the dominant position in the market. It has since become clear that, depending on the task and users, other input devices can significantly outperform the mouse (Milner, 1988). For example, when less than ten targets are displayed on screen and the cursor can be made to jump from one to the next, cursor keys are faster than a mouse (Shneiderman, 1987). In the electronic text domain, Ewing *et al*. (1986) found this to be the case with the HyperTIES application, though in this study the mouse seems to have been used on less than optimal surface conditions.

Though 'direct manipulation' (Shneiderman, 1984) might be a common description of an interface, it seems that its current manifestations leave much to be desired when it comes to manipulating text. Obviously practice and experience will play a considerable part here. Expertise with an input device affords the user a high level of control and breeds a sense of immediacy between selection and action.

It is important to realize that the whole issue of input device cannot be separated from other manipulation variables such as scrolling or paging through documents. For example, a mouse that must be used in conjunction with a menu for paging text will lead to different performance characteristics than one used with a scroll bar. On the basis of a literature review, Mills and Weldon (1986) report that there is no real performance difference between scrolling and paging electronic documents though Schwartz *et al*. (1983) found that novices tend to prefer paging (probably based on its close adherence to the book metaphor) and Dillon *et al*. (1990b) report that a scrolling mechanism was the most frequently cited improvement suggested by subjects assessing their reading interface.

For the moment, the mouse and scrolling mechanism appears dominant as the physical input and control device for text files and as the 'point and click' concept they embrace becomes integrated with the 'look and feel' of hypertext they will prove difficult to replace, even if convincing experimental evidence against their use, or an innovative credible alternative should emerge.[2]

3.7.4 Display size

Display size is a much discussed but infrequently studied aspect of human–computer interaction in general and reading electronic text in particular. Popular wisdom suggests that 'bigger is better' but empirical support for this edict is sparse. Duchnicky and Kolers (1983) investigated the effect of display size on reading constantly scrolling text and reported that there is little to be gained by increasing display size to more than four lines either in terms of reading speed or comprehension. Elkerton and Williges (1984) investigated 1, 7, 13, and 19-line displays and reported that there were few speed or accuracy advantages between the displays of seven or more lines. Similarly, Neal and Darnell (1984) report that there is little advantage in full page over partial page displays for text-editing tasks.

These results could be interpreted as suggesting that there is some critical point in display size, probably around five lines, above which improvements are slight. Intuitively this seems implausible. Few readers of paper texts would accept presentations of this format. Experiences with paper suggest that text should be displayed in larger units than this. Furthermore, loss of context is all too likely to occur with lengthy texts and the ability to browse and skim backward and forward is much easier with 30 or so lines of text than with 5-line displays (as we shall discuss later). Richardson *et al.* (1989) report a preference effect for 60-line over 30-line screens but no significant performance effect for an information location type task in a book length text so the jury is still out on the value of extremely large screens.

3.7.5 Conclusion on physical variables

There are wide ranging physical differences between paper and electronic documents that affect the manner in which readers use them. While myths abound on the theme of display size or orientation, research suggests that the major problems occur in designing appropriate manipulation facilities for the electronic medium. It is interesting to speculate how readers responses may alter as a function of familiarity and emerging standards for manipulation devices. To date though, the empirical literature would seem to indicate that the major source of the reported performance differences between the media does not lie at the physical level of reading.

3.8 Perceptual sources of difference

Most ergonomic work on reading from screens has concentrated on the perceptual differences between screen and paper presented text. With good reason, the image quality hypothesis emerged which held that many of the reported differences between the media resulted from the poorer quality of image available on screens. An exhaustive programme of work conducted by Gould and his colleagues at IBM between 1982 and 1987 represents probably the most rigorous and determined research effort. They tried to isolate a single variable responsible for observed differences. Reviews and summaries of this work have been published elsewhere (Gould *et al.*, 1987a; 1987b; Dillon, 1992) and will not be reproduced here. However several issues are worth raising in respect to this hypothesis of which developers of electronic documents need to be aware.

3.8.1 The image quality hypothesis

Characters are written on a VDU by an electron beam which scans the phosphor surface of the screen, causing stimulated sections to glow temporarily. The phosphor is characterized by its persistence, a high-persistence phosphor glowing for longer than a low-persistence phosphor. In order to generate a character that is apparently stable it is necessary to rescan the screen constantly with the requisite pattern of electrons. The frequency of scanning is referred to as the refresh rate.

As this brief description suggests, the quality of the image presented to the reader of an electronic document may be variable and unstable. This has led to the search for important determinants of image quality and their relationship to reader performance. Some of the major areas of attention have been flicker, screen dynamics, visual angle of view and image polarity.

Flicker

Since the characters are in effect repeatedly fading and being regenerated it is possible that they appear to flicker rather than remain constant. The amount of perceived flicker will obviously depend on both the refresh rate and the phosphor's persistence; the more frequent the refresh rate and the longer the persistence, the less perceived flicker. However refresh rate and phosphor persistence alone are not sufficient to predict whether or not flicker will be perceived by a user. It is also necessary to consider the luminance of the screen. While a 30 Hz refresh rate is sufficient to eliminate flicker at low luminance levels, Bauer *et al.* (1983) suggested that a refresh rate of 93 Hz is necessary in order for 99 per cent of users to perceive a display of dark characters on a light background as flicker free.

If flicker was responsible for the large differences between reading from paper and VDU it would be expected that studies such as those by Creed *et al.*

(1987) which employed photographs of screen displays would have demonstrated a significant difference between reading from photos and VDUs. However the extent to which flicker may have been an important variable in many studies is unknown as details of screen persistence and refresh rates are often not included in publications. Gould *et al.* (1987a) state that the photographs used in their study were of professional quality but appeared less clear than the actual screen display. It is likely that using photos to control flicker may not be a suitable method and flicker may play some part in explaining the differences between the two media.

Screen dynamics

In order to understand the role of such dynamic variables in reading from VDUs, Gould *et al.* (1987a) had subjects read from paper, VDU and good quality photographs of the VDU material which maintained the screen image but eliminated any possible dynamics. Results provided little in the way of firm evidence to support the idea of dynamics causing problems. Creed *et al.* (1987) also compared paper, VDU and photos of the screen display on a proof-reading task with thirty subjects. They found that performance was poorest on VDU but photographs did not differ significantly from either paper or VDU in terms of speed or accuracy, though examination of the raw data suggested a trend towards poorer performance on photos than paper. It seems unlikely therefore that much of the cause for differences between the two media can be attributed to the dynamic nature of the screen image.

Visual angle and viewing distance

Gould and Grischkowsky (1986) hypothesized that due to the usually longer line lengths on VDUs the visual angle subtended by lines in each medium differs and that people have learned to compensate for the longer lines on VDUs by sitting further away from them when reading. Gould and Grischkowsky (1986) had 18 subjects read twelve different three-page articles for misspellings. Subjects read two articles at each of six visual angles: 6.7, 10.6, 16.0, 24.3, 36.4 and 53.4 degrees, varied by maintaining a constant reading distance while manipulating the image size used. Results showed that visual angle significantly affected speed and accuracy. However the effects were only noticeable for extreme angles, and between a range of 16.0 to 36.4 degrees, which covers typical VDU viewing, no effect for angle was found. Brand and Judd (1993) report that visual angle of hard-copy in text editing tasks effected performance in a manner largely consistent with the findings of Gould and Grischkowsky (1986). Other work has confirmed that preferred viewing distance for screens is greater than that for paper (Jaschinski-Kruza, 1990).

Image polarity

A display in which dark characters appear on a light background (e.g. black on white) is referred to as positive image polarity or negative contrast. This will be referred to here as positive presentation. A display on which light characters appear on a dark background (e.g. white on black) is referred to as negative image polarity or positive contrast. This will be referred to here as negative presentation. The traditional computer display involves negative presentation, typically white on black though light green on dark green is also common.

Since 1980 there has been a succession of publications concerned with the relative merits of negative and positive presentation. Several studies suggest that, tradition notwithstanding, positive presentation may be preferable to negative. For example, Radl (1980) reported increased performance on a data input task for dark characters and Bauer and Cavonius (1980) reported a superiority of dark characters on various measures of typing performance and operator preference.

With regard to reading from screens Cushman (1986) reported that reading speed and comprehension on screens was unaffected by polarity, though there was a non-significant tendency for faster reading of positive presentation. Gould *et al.* (1987a) specifically investigated the polarity issue. Fifteen subjects read five different 1000 word articles: two were negatively presented, two positively presented and one on paper (standard positive presentation). Further experimental control was introduced by fixing the display contrast for one article of each polarity at a contrast ratio of 10:1 and allowing the subject to adjust the other article to their own liking. This avoided the possibility that contrast ratios may have been set which favoured one display polarity. Results showed no significant effect for polarity or contrast settings, though 12 of the 15 subjects did read faster from positively presented screens, leading the investigators to conclude that display polarity probably accounted for some of the observed differences in reading from screens and paper.

In a general discussion of display polarity Gould *et al.* (1987b) state that 'to the extent that polarity makes a difference it favours faster reading from dark characters on a light background' (p. 514). Furthermore they cite Tinker (1963) who reported that polarity interacted with type size and font when reading from paper. The findings of Bauer *et al.* (1983) with respect to flicker certainly indicate how perceived flicker can be related to polarity. Therefore the contribution of display polarity in reading from screens is probably important through its interactive effects with other display variables.

Display characteristics

Issues related to fonts such as character size, line spacing and character spacing have been subjected to detailed research. However the relationship of much of the findings to reading continuous text from screens is not clear.

Character size on VDUs is closely related to the dimension of the dot-matrix from which the characters are formed. The dramatic increase in computer processing power now means that there is little cost in employing larger matrices and Cakir *et al.* (1980) recommend a minimum of 7×9. Pastoor *et al.* (1983) studied the relative suitability of four different dot-matrix sizes and found reading speed varied considerably. On the basis of these results the authors recommended a 9×13 character size matrix.

Considerable experimental evidence exists to favour proportionally rather than non-proportionally spaced characters (e.g. Beldie *et al.*, 1983). Once more though, the findings must be viewed cautiously. In the Beldie *et al.* study for example, the experimental tasks did not include reading continuous text. Muter *et al.* (1982) compared reading speeds for text displayed with proportional or non-proportional spacing and found no effect. In an experiment intended to identify the possible effect of such font characteristics on the performance differences between paper and screen reading, Gould *et al.* (1987a) also found no evidence to support the case for proportionally spaced text.

Kolers *et al.* (1981) studied interline spacing and found that with single spacing significantly more fixations were required per line, fewer lines were read and the total reading time increased. However the differences were small and were regarded as not having any practical significance. On the other hand Kruk and Muter (1984) found that single spacing produced 10.9 per cent slower reading than double spacing, a not inconsiderable difference.

Muter and Maurutto (1991) attempted various 'enhancements' to screen presented text to see if they could improve reading performance. These included double spacing between lines, proportional spacing within words, left justification only and positive presentation. 'Enhanced' text proved to be read no differently from more typical electronic text (i.e. basically similar to paper) which the authors state may be due to one or two of their 'enhancements' having a negative and therefore neutralizing effect on others or some 'enhancements' interacting negatively. Unfortunately, their failure to manipulate such variables systematically means firm conclusions cannot be drawn.

Obviously much work needs to be done before a full understanding of the relative advantages and disadvantages of particular formats and types of display is achieved. In a discussion of the role of display fonts in explaining any of the observed differences between screen and paper reading Gould *et al.* (1987a) conclude that font has little effect on reading rate from paper (as long as the fonts tested are reasonable). They add that it is almost impossible however to discuss fonts without recourse to the physical variables of the computer screen itself, e.g. screen resolution and beam size, once more highlighting the potential cumulative effect of several interacting factors on reading from screens.

Anti-aliasing

Most computer displays are raster displays typically containing dot-matrix characters and lines which give the appearance of 'staircasing', i.e. edges of characters may appear jagged. This is caused by undersampling the signal that would be required to produce sharp, continuous characters. The process of anti-aliasing has the effect of perceptually eliminating this phenomenon on raster displays. A technique for anti-aliasing developed by IBM accomplishes this by adding variations in grey level to each character.

The advantage of anti-aliasing lies in the fact that it improves the quality of the image on screen and facilitates the use of fonts more typical of those found on printed paper. To date the only reported investigation of the effects of this technique on reading from screens is that of Gould *et al.* (1987b). They had 15 subjects read three different 1000 word articles, one on paper, one on VDU with anti-aliased characters and one on VDU without anti-aliased characters. Results indicated that reading from anti-aliased characters did not differ significantly from either paper or aliased characters though the latter two differed significantly from each other. Although the trend was present the results were not conclusive and no certain evidence for the effect of anti-aliasing was provided. However the authors report that 14 of the 15 subjects preferred the anti-aliased characters, describing them as clearer and easier to read.

3.8.2 The interaction of display variables and the effect of image quality

Despite many of the findings reported thus far, it appears that reading from screens can at least be as fast and as accurate as reading from paper. Gould *et al.* (1987b) have empirically demonstrated that under the right conditions and for a strictly controlled proof-reading task, such differences between the two presentation media disappear. In a study employing 16 subjects, an attempt was made to produce a screen image that closely resembled the paper image, i.e. similar font, size, colouring, polarity and layout were used. Univers-65 font was positively presented on a monochrome IBM 5080 display with an addressability of 1024×1024. No significant differences were observed between paper and screen reading. This study was replicated with 12 further subjects using a 5080 display with an improved refresh rate (60 Hz). Again no significant differences were observed though several subjects still reported some perception of flicker.

On balance it appears that any explanation of these results must be based on the interactive effects of several of the variables outlined in the previous sections. After a series of experimental manipulations aimed at identifying those variables responsible for the improved performance Gould *et al.* (1987b) suggested that the performance deficit was the product of an interaction between a number of individually non-significant effects. Specifically, they identified display polarity (dark characters on a light, whitish background), improved display resolution, and anti-aliasing as major contributions to the elimination of the paper/screen reading rate difference.

Gould *et al.* (1987b) conclude that the explanation of many of the reported differences between the media is basically visual rather than cognitive and lies in the fact that reading requires discrimination of characters and words from a background. The better the image quality is, the more reading from screen resembles reading from paper and hence the performance differences disappear. This seems an intuitively sensible conclusion to draw. It reduces to the level of simplicity any claims that one or other variable such as critical flicker frequency, font or polarity is responsible for any differences. As technology improves we can expect to see fewer speed deficits at least for reading from screens. Recent evidence from Muter and Maurutto (1991) using a commercially available screen has shown this to be the case, although other differences remain.

A major shortcoming of the studies by Gould *et al.* is that they only address limited outcome variables: speed and accuracy. Obviously speed is not always a relevant criterion in assessing the output of a reading task. Furthermore, the accuracy measures taken in these studies have been criticized as too limited and further work needs to be carried out to appreciate the extent to which the explanation offered by Gould is sufficient. It follows that other observed outcome differences such as fatigue, reader preference and comprehension should also be subjected to investigation in order to understand how far the image quality hypothesis can be pushed as an explanation for reading differences between the two media.

A shortcoming of most work cited in this section is the task employed. Invariably it was proof-reading which hardly constitutes normal reading for most people. Thus the ecological validity of many of these studies is low. Beyond this, the actual texts employed were all relatively short (Gould's for example averaged only 1100 words and many other researchers used even shorter texts). As a result, it is difficult to generalize these conclusions beyond the specifics of task and texts employed to the wider class of activities termed 'reading'. Creed *et al.* (1987) defend the use of proof-reading on the grounds of its amenability to manipulation and control. While this desire for experimental rigour is laudable, one cannot but feel that the major issues involved in using screens for real-world reading scenarios are not addressed by such work.

3.9 Cognitive sources of difference

It is clear that the search for the specific ergonomic variables responsible for differences between the media has been insightful. However, few readers of electronic texts would be satisfied with the statement that the differences between the media are physical or visual rather than cognitive. This might explain absolute speed and accuracy differences on linked tasks but hardly accounts for the range of process differences that are observed or experienced. Once the document becomes too large to display on a single screen other factors than image immediately come into play. At this stage readers must start to manipulate the document and thus be able to relate current to previously-displayed material. In such a situation other factors such as memory for text and its location, ability to search for items and speed of movement through the document come into play and the case for image quality as the major determinant of performance is less easy to sustain. Therefore in this section, other possible factors at the cognitive level are discussed.

3.9.1 Short-term memory for text

A related issue to display size and scrolling/paging mentioned earlier is the splitting of paragraphs midsentence across successive screens. In this case, which is more likely to occur in small displays, the reader must manipulate the document in order to complete the sentence. This is not a major issue for paper texts such as books or journals because the reader is usually presented with two pages at a time and access to previous pages is normally easy. On screen however, access rates are not so fast and the break between screens of text is likely to be more critical.

Research into reading has clearly demonstrated the complexity of the cognitive processing that occurs. The reader does not simply scan and recognize every letter in order to extract the meaning of words and then sentences. Comprehension is thought to require inference and deduction, and skilled readers probably achieve much of their smoothness by predicting probable word sequences (Chapman and Hoffman, 1977; though see Mitchell, 1982). The basic units of comprehension in reading that have been proposed are propositions (Kintsch, 1974), sentences (Just and Carpenter, 1980) and paragraphs (Mandler and Johnson, 1977). Splitting sentences across screens is likely to disrupt the process of comprehension by placing an extra burden on the limited capacity of working memory to hold the sense of the current conceptual unit while the screen is filled. Furthermore, the fact that 10–20 per cent of eye movements in reading are regressions to earlier fixated words and that significant eye-movement pauses occur at sentence ends (Ellis, 1984) would suggest that sentence splitting is also likely to disrupt the reading process and thereby hinder comprehension.

Dillon *et al.* (1990b) found that splitting text across screens caused readers to return to the previous page to re-read text significantly more often than when text was not split. Though this appeared to have no effect on subsequent comprehension of the material being read, they concluded that it was remarked upon by the subjects sufficiently often to suggest that it would be a nuisance to regular users. In this study however the subjects were reading from a paging rather than scrolling interface where the effect of text splitting was more likely to cause problems due to screen-fill delays. With scrolling interfaces text is always going to split across screen boundaries but there is rarely a perceptible delay in image presentation to disrupt the reader. It would seem therefore that to the extent to which such effects are likely to be noticeable, text splitting should be avoided for paging interfaces.

3.9.2 Visual memory for location

There is evidence to suggest that readers establish a visual memory for the location of items within a printed text based on their spatial location both on the page and within the document (Rothkopf, 1971; Lovelace and Southall, 1983). This memory is supported by the fixed relationship between an item and its position on a given page. A scrolling facility is therefore liable to weaken these relationships and offers the reader only the relative positional cues that an item has with its immediate neighbours.

It has become increasingly common to present information on computer screen via windows, i.e. sections of screen devoted to specific groupings of material. Current technology supports the provision of independent processes within windows or the linking of inputs in one window with the subsequent display in another, the so-called 'co-ordinated windows' approach (Shneiderman, 1987).

Such techniques have implications for the presentation of text on screen as they provide alternatives to the straightforward listing of material in 'scroll' form or as a set of 'pages'. For example, while one window might present a list of contents in an electronic text, another might display whole sections of it according to the selection made. In this way, not only is speed of manipulation increased but the reader can be provided with an overview of the document's structure to aid orientation while reading an opened section.

The use of such techniques is now commonplace in hypertext applications. GUIDE for example, uses windows in one instance to present short notes or diagrams as elaborations or explanations of points raised in the currently viewed text, rather like sophisticated footnotes. The concept of hypertext as non-linear text is, in a very real sense, derived from such presentation facilities.

Tombaugh *et al.* (1987) investigated the value of windowing for readers of lengthy electronic texts. They had subjects read two texts on single or multi-window formats before performing ten information location tasks. They found that novices initially performed better with a single-window format but subsequently observed that, once familiar with the manipulation facilities, the benefits of multi-windowing in terms of aiding spatial memory became apparent. They highlight the importance of readers acquiring familiarity with a system and the concept of the electronic book in order to accrue the benefits of such facilities.

Simpson (1989) compared performance with a similar multi-window display, a 'tiled' display (in which the contents of each window were permanently visible) and a 'conventional' stack of windows (in which the windows remained in reverse order of opening). She reported that performance with the conventional window stack was poorest but that there was no significant difference between the 'tiled' and multi-window displays. She concluded that for information location tasks, the ability to see a window's contents is not as important as being able to identify a permanent location for a section of text.

Stark (1990) asked people to examine a hypertext document in order to identify appropriate information for an imaginary client and manipulated the scenario so that readers had to access information presented either in a 'pop-up' window which appeared in the top right hand corner of the screen or a 'replacement' window which overlaid the information currently being read. Though no significant task performance or navigation effects were observed, subjects seemed more satisfied with pop-ups than replacements.

Such studies highlight the impact of display format on readers' performance of a standard reading task: information location. Spatial memory seems important and paper texts are good at supporting its use through permanence of format. Windowing, if deployed so as to retain order can be a useful means of overcoming this inherent weakness of electronic text. However, studies examining the problems of windowing very long texts (where more than five or six stacked windows or more frequent window manipulations are required) need to be performed before any firm conclusions about the benefits of this technique can be drawn. (See e.g. Jones and Dumais, 1986.)

3.9.3 Schematic representations of documents

Exposure to the variety of texts in everyday life could, according to contemporary cognitive theories, lead readers to acquire mental models or schemata for documents with which they are familiar. Schemata are hypothetical knowledge structures which mentally represent the general attributes of a concept and are considered by psychologists to provide humans with powerful organizing principles for information (Bartlett, 1932; Cohen, 1988). Thus, when we pick up a book we immediately have expectations about the likely contents. Inside the front cover we expect such details as where and when it was published, perhaps a dedication and then a Contents page. We know, for example, that contents listings describe the layout of the book in terms of chapters, proceeding from the front to the back. Chapters are organized around themes and an index at the back of the book, organized alphabetically, provides more specific information on where information is located in the body of the text. Experienced readers know all this before even opening the text. It would strike us as odd if such structures were absent or their positions within the text were altered.

According to van Dijk and Kintsch (1983), such models or schemata, which they term 'superstructures', facilitate comprehension of material by allowing readers to predict the likely ordering and grouping of constituent elements of a body of text. To quote van Dijk (1980, p. 108): 'a superstructure is the schematic form that organizes the global meaning of a text. We assume that such a superstructure consists of functional categories . . . (and) . . . rules that specify which category may follow or combine with what other categories.'

But apart from categories and functional rules, van Dijk adds that a superstructure must be socioculturally accepted, learned, used and commented upon by most adult language users of a speech community. Research by van Dijk and Kintsch (1983) and Kintsch and Yarborough (1982) has shown how such structures influence comprehension of texts.

In this format the schema/superstructure constitutes a set of expectancies about their usual contents and how they are grouped and positioned relative to each other. In advance of actually reading the text readers cannot have much insight into anything more specific than this, but the generality of organization within the multitude of texts read in everyday life affords stability and orientation in what could otherwise be a complex informational environment.

The concept of a schema for an electronic information space is less clear-cut than those for paper documents. Electronic documents have a far shorter history than paper and the level of awareness of technology among the general public is relatively primitive compared with that of paper. Exposure to information technology will almost certainly improve this state of affairs but even among the contemporary computer literate it is unlikely that the type of generic schematic structures that exist for paper documents have electronic equivalents of sufficient generality.

Obviously computing technology's short history is one of the reasons but it is also the case that the medium's underlying structures do not have equivalent transparency. Thus using electronic information is often likely to involve the employment of schemata for systems in general (i.e. how to operate them) in a way that is not essential for paper-based information.

The qualitative differences between the schemata for paper and electronic documents can easily be appreciated by considering what you can tell about either at first glance. The information available to paper text users was outlined above. When we open a hypertext or other electronic document however we do not have the same amount of information available to us. We are likely to be faced with a welcoming screen which might give us a rough idea of the contents (i.e. subject matter) and information about the authors/developers of the document but little else. Such displays are usually two-dimensional, give no indication of size, quality of contents, age (unless explicitly stated) or how frequently the text has been used (i.e. there is no dust or signs of wear and tear on it such as grubby finger-marks or underlines and scribbled comments).

Performing the electronic equivalent of opening up the text or turning the page offers no assurance that expectations will be met. Many hypertext documents offer unique structures (intentionally or otherwise) and their overall sizes are often impossible to assess in a meaningful manner; these points are dealt with in more detail by Dillon *et al.* (1993). At their current stage of development it is likely that users/readers familiar with hypertext will have a schema that includes such attributes as linked nodes of information, non-serial structures, and perhaps, potential navigational difficulties! The manipulation facilities and access mechanisms available in hypertext will probably occupy a more prominent role in their schemata for hypertext documents than they will for readers' schemata of paper texts. As yet, empirical evidence for such schemata is lacking.

The fact that hypertext offers authors the chance to create numerous structures out of the same information is a further source of difficulty for users or readers. Since schemata are generic abstractions representing typicality in entities or events, the increased variance of hypertext implies that any similarities that are perceived must be at a higher level or must be more numerous that the schemata that exist for paper texts.

It seems therefore that users' schemata of electronic texts are likely to be 'informationally leaner' than those for paper documents. This is attributable to the recent emergence of electronic documents and comparative lack of experience interacting with them as opposed to paper texts for even the most dedicated users. The lack of standards in the electronic domain compared to the rather traditional structures of many paper documents and the paucity of affordances or cues to use are further problems for schema development with contemporary electronic texts.

3.9.4 Searching

Electronic text supports word or term searches at rapid speed and with total accuracy and this is clearly an advantage for users in many reading scenarios, e.g. checking references, seeking relevant sections, etc. Indeed it is possible for such facilities to support tasks that would place unreasonable demands on users of paper texts, e.g. searching a large book for a non-indexed term or several volumes of journals for references to a concept.

Typical search facilities require the user to input a search string and choose several criteria for the search such as ignoring certain text forms (e.g. all uppercase words) but sophisticated facilities on some database systems can support specification of a range of texts to search. The usual form for search specification is Boolean, i.e. users must input search criteria according to formal rules of logic employing the constructs 'either', 'or' as well as 'and', which when used in combination support powerful and precise specifications. Unfortunately most end-users of computer systems are not trained in their use and while the terms may appear intuitive, they are often difficult to employ successfully.

In current electronic text facilities a simple word search is common but users still seem to have difficulties. Richardson *et al.* (1988) reported that several subjects in their experiment displayed a tendency to respond to unsuccessful searches by increasing the specificity of the search string rather than lessening it. The logic appeared to be that the computer required precision rather than approximation to search effectively. While it is likely that such behaviour is reduced with increased experience of computerized searching, a study by McKnight *et al.* (1990a) of information location within text found other problems. Here, when searching for the term 'wormwood' in an article on wine making, two subjects input the search term 'woodworm', displaying the intrusion of a common-sense term for an unusual word of similar sound and shape (a not uncommon error in reading under pressure due to the predictive nature of this act during sentence processing). When the system correctly returned a 'Not Found' message, both users concluded that the system could not fail, hence the question was an experimental trick.

Thus it seems as if search facilities are a powerful means of manipulating and locating information on screen and convey certain advantages impossible to provide in the paper medium. However, users may have difficulties with them in terms of formulating accurate search criteria. This is an area where research into the design of search facilities and increased exposure of users to electronic information can lead to improvements resulting in a positive advantage of electronic text over paper.

3.9.5 Individual differences in skill, knowledge and experience

It has been noted that many of the studies reported in this review employed relatively naïve users as subjects. The fact that different types of users interact with computer systems in different ways has long been recognized and it is possible that the differences in reading that have been observed in these studies result from particular characteristics of the user group involved.

Most obviously, it might be assumed that increased experience in reading from computers would reduce the performance deficits. A direct comparison of experienced and inexperienced users was incorporated into a study on proof-reading from VDUs by Gould *et al.* (1987a). Experienced users were described as 'heavy, daily users . . . and had been so for years'. Inexperienced users had no experience of reading from computers. No significant differences were found between these groups, both reading slower from screen.

Smedshammar *et al.* (1989) report that *post hoc* analysis of their data indicate that fast readers are more adversely affected by VDU presentation than slow readers. However, their classification of reading speed is based on mean performance over three conditions in their experiment rather than controlled, pre-trial selection suggesting caution in drawing conclusions. Smith and Savory (1989) report an interaction effect between presentation medium, reading strategy and susceptibility to external stress measured by questionnaire suggesting that working with VDUs may exaggerate some differences in reading strategy for individuals with high stress levels. Caution in interpretation of these results is suggested by the authors.

No reported differences for age or sex can be found in the literature. Therefore it seems reasonable to conclude that basic characteristics of the user are not responsible for the differences in reading from these presentation media.

3.10 General conclusion: so what do we know now?

At the outset it was stated that reading can be assessed in terms of outcome and process measures. To date however, most experimental work has concentrated on the former and in particular, has been driven by a desire to identify a single variable to account for the significant reading speed differences that have been reported. The present review sought to examine the experimental literature with a view to identifying all relevant issues and show how single variable explanations are unlikely to offer a satisfactory answer.

While substantial progress has been made in terms of understanding the impact of image quality on reading speed, it is clear that ergonomics are still a long way from understanding fully the effect of presentation medium on reading. While it is now possible to draw up recommendations on how to ensure no speed deficit for proof-reading short texts on screen, changes in task and text parameters mean such advice has less relevance.

One is struck in reviewing this literature by the rather limited view of reading that some investigators seem to have. Most seem to concern themselves with the control of so many variables that the resulting experimental task bears little resemblance to the activities most of us routinely perform under the banner 'reading'. It is perhaps no coincidence that the major stumbling block of reader preference has been so poorly investigated beyond the quick rating of screens and test documents in post-experimental surveys.

The assumption that overcoming speed or accuracy differences in proof-reading is sufficient to claim, as some authors have, that 'there is no difference' between the media (Oborne and Holton, 1988) is testimony to the limitations of some ergonomists' views of human activities such as reading. Other tasks, such as reading to comprehend, to learn or for entertainment are less likely to require readers to concern themselves with speed. These are the sort of tasks people will regularly wish to perform and it is important to know how electronic text can be designed to support them. Such tasks will also of necessity involve a wide variety of texts, differing in length, detail, content-type and so forth – issues that have barely been touched upon to date by researchers.

The findings on image quality and the emerging knowledge of manipulation problems should not be played down however. Knowing what makes for efficient visual processing and control of electronic text can serve as a basis for future applications. As Muter and Maurutto (1991) demonstrated, a typical high quality screen with effective manipulation facilities can provide an environment that holds its own in speed, comprehension and preference terms with paper, at least over the relatively constrained reading scenarios found in the researchers' laboratory. But if our desire is to create systems that improve on paper rather than just matching it in performance and satisfaction terms (as it should be) then much more work and a more realistic conceptualization of human reading is required.

Notes

1. As Nelson's book is distributed as a hypertext document there are no page numbers. However, this quote can be located in Chapter 1, An Obvious Vision, under the heading Tomorrow's World of Text on Screen. Such lengthy reference to a specific section highlights a potential problem with hypertext that must be addressed.
2. One has only to consider the dominance of the far from optimal QWERTY keyboard to understand how powerful convention is. This keyboard was actually designed to slow human operators down by increasing finger travel distances thereby lessening key jamming in early mechanical designs. Despite evidence that other layouts can provide faster typing speeds with less errors (Martin 1972) the QWERTY format retains its dominant position.

4

Describing the reading process at an appropriate level

It is a good morning exercise for a research scientist to discard a pet hypothesis everyday before breakfast. It keeps him young.

Konrad Lorenz, *On Aggression* (1963)

4.1 Introduction

Taken in isolation, the empirical literature on reading from screens versus paper is not particularly helpful to a designer or design team concerned with developing an electronic text application. The classification of the empirical literature provided in the previous chapter is an attempt to afford better conceptualization of the relevant issues and experimental findings but it suffers from the problems inherent in the literature itself: the absence of a suitable descriptive framework of the reader that would enable those concerned with electronic text to derive guidance for specific design applications.

As it stands, the empirical literature presents two implicit views of the typical reader and provides recommendations accordingly. The first is as a scanner of short texts, searching out spelling mistakes or some relatively trivial error, often in a race against the clock (and usually trying to impress an experimenter). The second is as a navigator travelling through a maze of information. These extremes are tempered only slightly by concessions to text or task variables as influences on the reading process and it is rare that any attempt to place reading in a broader, more realistic context is made. Yet reading is a form of information usage that rarely occurs as a self-contained process with goals expressible in terms of speed or number of items located. Far more frequently it occurs as a means to, or in support of, non-trivial everyday ends such as keeping informed of developments at work, checking bank balances, understanding how something works (or more likely, why right now it is not working) and so forth. These are precisely the type of information acts, of which reading is an essential component, that people perform routinely. The success of these acts is measured in effects not usually quantifiable simply in terms of time taken or errors made. True, readers do notice spelling mistakes, they may even proof-read as professionals and they certainly must manipulate and navigate through lengthy documents, but such views alone can never adequately describe the situated realities and totality of the reading situation.

This constrained view of the reading process becomes even more apparent when one examines the conceptualizations of reading that dominate the various disciplines which lay claim to some interest in it. Psychology, a discipline that might justifiably consider itself directly concerned with understanding reading is, according to Samuels and Kamil (1984), concerned with 'the *entire* process from the time the eye meets the page until the reader experiences the "click of comprehension"' (p. 185), italics added).

This sounds suitably all-embracing but in reality is relatively narrow when one realizes the everyday attributes of reading that it overlooks. There are few psychological models of reading that consider text manipulation or navigation for example as part of the reading process. The literature which provides theoretical input to these domains is usually the product of other research issues such as memory organization or word learning. Can any view that claims reading only starts when the 'eye meets the page' really lay claim to covering the 'entire' process? Certainly it appears logical to start here but if the situation demands that the readers moves from text to text in their search for information do we conclude that the intervening moves are not part of the process of reading? And if readers fails to experience that 'click of comprehension' has the reading process stopped at some point when the eye leaves the text or does it extend while they think the material through and get that 'click' later? Interestingly, not all psychologists would even accept Samuels and Kamil's definition of psychology's legitimate concerns with reading. Crowder (1982) for example, explicitly excludes comprehension as an issue in his analysis of the reading process.

Rumelhart's (1977) theory of reading marked a supposed breakthrough in cognitive models of reading by highlighting the limits of linear models (those proposing a one-way sequence of processing stages from visual input to comprehension). He outlined an alternative interactive form (one supporting the influence of higher stages of processing on lower ones) which accounted for some of the experimental findings that were difficult to accommodate in linear models. His model, parts of which have been successfully implemented in software form thereby passing possibly the strongest test of a contemporary cognitive theory (McClelland and Rumelhart, 1981), represents reading largely as an act of word recognition. The model has been summarized as follows by de Beaugrande (1981, p. 281).

> Reading begins with the recognition of visual features in letter arrays. A short lived iconic image is created in brief sensory storage and scanned for critical determinants. Available features are fed as oriented line segments into a pattern synthesizer that, as soon as it is confident about what image has been detected, outputs an abstract characterisation . . . The extracted

features are constraints rather than determiners, interacting with context and reader expectations . . . The individual letters are heavily anticipated by stored representations in a 'word index'. Even in recognition of letters and words all of the various sources of knowledge, both sensory and non-sensory, come together in one place and the reading process is the product of the simultaneous joint application of all knowledge sources.

From such a description it is not difficult to understand why Venezky (1984, p. 4) states that: 'the history of research on the reading process is for the most part the history of cognitive psychology' and, one might add, the reason why cognitive psychology is often accused of an under-critical acceptance of the computer metaphor of mind. However, it strongly emphasizes the limitations of such work for system designers. There is no mention of a reading task, a text, a goal, or a context in which these processes occur. 'Visual feature recognition', 'iconic images' and 'pattern synthesizers' are theoretical constructs which attempt to provide a plausible account of how humans extract information from text (or 'letter arrays' to use the jargon) but mapping findings from such analyses and models to the world of electronic text design is, by definition, beyond the model's scope.

The narrow focus of much reading research is reflected sharply in the opening pages of psycholinguist Frank Smith's (1978) book titled *Reading* where he remarks that a glance through the text might leave one justifiably thinking that 'despite its title, this book contains little that is specifically about reading' (p. 1).

He goes on the argue for the need to examine reading in a reductionist manner (claiming that there is little about reading that is unique – it involves certain cognitive processes and structures that researchers not interested in reading have already investigated in other contexts – presuming that such work transfers appropriately to discussions of the reading process which is itself a debatable assumption), but it is his early admission of the lack of real-life relevance of the work that produced the most memorable 'click of comprehension' in this reader's mind.

A recent and vociferous attack on the standard cognitive psychological approach to reading comes from Kline (1988, p. 36). In a book given over to attacking much of contemporary experimental psychology, he singles out reading as a prime example of the lack of validity in much of the discipline's work. Describing a typical reading experiment investigating people's categorization of sentences as meaningful or non-meaningful while their reaction time is measured, he states:

> the task . . . is not really like reading. Certainly it involves reading but most people read for pleasure or to gain information. Furthermore, reading has serious emotional connotations on occasion, as for example reading pornography, a letter informing you that your PhD has been turned down or your lover is pregnant . . . Furthermore, most adults, when reading books especially, read large chunks at a time.

Kline continues, humorously comparing lines from a Shakespeare sonnet (e.g.

'Like as the waves make towards the pebbled shore, so do our minutes hasten to their end') with lines from such experimental tasks (e.g. 'Canaries have wings – true or false?'; 'Canaries have gills – true or false?') before concluding that such work is absurd in the context of real reading and its resultant theoretical models of no predictive and little explanatory use. His criticisms might seem harsh and populist were it not for the fact that Kline is a psychologist of international reputation and admits to having performed, like most psychologists, such experiments himself earlier in his career.

But psychology is not unique in failing to provide a satisfactory account of the process for the purposes of design. Information science, the contemporary theoretical backbone of librarianship might also be viewed as having a natural interest in the reading process. Yet its literature offers few clues to those concerned with designing electronic texts for reading. As Hatt (1976, p. 3) puts it:

> A great body of professional expertise has been developed, the general aims of which have been to improve the provision of books and to facilitate readers' access to books. At the point where the reader and the book come together however, it has been the librarian's habit to leave the happy pair and tiptoe quietly away, like a Victorian novelist.

Hatt argues that the problems of information science are the problems of all disciplines concerned with this subject and that although much valuable work has been done and knowledge has been gained, he draws a similar (though less emphatically expressed) conclusion to Kline in that one comes away from the literature thinking 'that's all well and good, but it's not really reading!'

In defence of each discipline it must be said that their approaches and methods reflect their aims. If psychology really is concerned with what happens between the moments when the eye meets the page and the reader understands the text (or just before that in the case of Crowder *inter alia*), then models of eye movements and word recognition have a place, despite Kline's enthusiastic dismissal. Though one might add that if this is all many psychologists consider important in understanding the reading process then Kline might really have a point. Few, if any, theorists interested in reading claim to cover all issues. What is pertinent here however is the irrelevance of much of this work to the issues associated with the development of more usable electronic documents, the depressing fact that much excellent work in psychology and allied sciences fails to provide much by way of influence in practical software design settings.

The unsuitability of any theoretical description of reading is a major (perhaps even *the*) problem for human factors work. Viewing the reader as an 'information processor' or a 'library user' and the reading process as a 'psycholinguistic guessing game' depending on theoretical stance, hardly affords prescriptive measures for the design of electronic text systems. The typical reader can certainly be said to process information and occasionally use libraries, but each is only a small part of the whole that is reading. If one deals exclusively with such limited aspects (as many theories do) the broad picture never emerges and this gives rise to the type of limited findings on text design one finds in much of the human factors literature.

4.2 The problem of theoretical description for human factors work

The problem for human factors work induced by incomplete descriptions is epitomized in a case study of a commercial system involving the author (Dillon, 1988a and Dillon *et al.*, 1991). A publishing consortium recently funded the development of an experimental system to support the document supply industry. Named ADONIS, the resulting workstation is designed to facilitate searching, viewing and printing of CD-ROM stored articles. It boasts a high-resolution A4-size screen that presents bit-mapped reproductions of journal articles. The trial system presented users with access to biomedical journals (selected on the basis of a usage study) and the workstation was aimed primarily at document supply staff working in storage centres who would process inter-library loan requests for articles, though possibilities for end-users in libraries to use the system directly existed and were considered by the development consortium as worthy of investigation (i.e. further stakeholders were identified by the developers and proposed as potential end-users). Thus the system came to be seen as a prototype electronic text system of the future.

The author was asked to evaluate the system from a human factors perspective for both user types. ADONIS's application to the document supply industry was a relatively straightforward evaluation and will not be discussed further here except to emphasize that it became clear from the task and user analyses carried out on-site that these users did not view documents with the system and under no circumstances could be described as readers of the material contained in the database. The potential end-users in libraries on the other hand were typical readers of journal articles and the evaluation in this context highlighted the shortcomings in current knowledge of electronic text design.

The specific details of the evaluation are not important for the present discussion (they are presented in the references cited). Time allowed for this work was limited as the developers were not keen on lending the equipment to remote sites for extended periods of time thus necessitating a quick rather than formal experimental approach on the part of the author. Suffice to say that the evaluation involved a cognitive walk-through performed by the present author, and the setting of three tasks for a sample of ten users to perform in an informal manner, i.e. with the evaluator present and the participants commenting on the system's user interface as they worked through the tasks. The tasks were so designed to ensure that readers were exposed to all aspects and functions of the ADONIS interface. Measures of speed and accuracy were eschewed in favour of general ratings of the system and comments on good or bad aspects of the interface which the evaluator noted as the subjects proceeded. As a result of frequent criticisms of the search facilities, a survey of normal procedures for citing articles among 35 researchers was also carried out to identify a more suitable set and layout for the search fields. The results of the evaluations were summarized, related to the literature on electronic text and general interface design and presented to the publishing consortium in the form of a written report.

Superficially, ADONIS was a good design. The large, high quality screen presented articles in an easy to read manner that conformed precisely to the structure of the paper version. By using a large screen and positive presentation it even adhered to some of the human factors design guidelines in the literature. Use of menus and a form-filling screen for inputting search criteria should have removed any learning difficulties for novice users too. The ability to store and retrieve a large number of articles from one system coupled with the ability to view material on screen before deciding whether or not to print it out, seemed to convey benefits to the ADONIS workstation not even possessed by paper, never mind other databases.

The readers studied at HUSAT however were very critical of the system. Common criticisms related to the style of the ADONIS interface which was described frequently as 'archaic', the speed of searching which was perceived as far too slow, the inability to search on keywords, and the restricted manipulation facilities available when an article was being viewed (ADONIS by virtue of its reliance on bit-mapped images was slow and only supported paging forwards or backwards, jumping directly to particular sections was impossible).

In an attempt to understand the likelihood of potential users actually reading texts with ADONIS, they were asked to comment on the readability of the displayed document independently of the manipulation facilities offered. Only two users said they would read with it; of the remainder, six said they would only scan articles prior to printing them out and two said they would never use it. In other words, although the system was designed in partial accordance with the literature on electronic text, it was rejected by these potential users. How could this be?

What was shown by the evaluation is that while ADONIS supports the ends, it fails to adequately provide the means. In other words, it emphasized functionality at the expense of usability, letting users get *what* they want but not *how* they want to. Users can obtain hardcopies of journal articles but they must master the counter-intuitive specification form first.[1] They can browse articles on a high quality large screen, but they cannot manipulate pages with the ease they can with paper. They can search an equivalent of a library of journals from their desk to obtain an article they seek, but they cannot browse through a list of contents and serendipitously discover a relevant title or author as they can with paper.

This clash between means and ends provides an interesting insight into the problems faced by many designers of electronic text systems (or indeed, information systems in general) which will be referred to here as the 'levels of description' problem. Briefly stated, it implies that there are various levels of abstraction at which human behaviour can be described and, while each may be accurate in itself, there exists an optimum level for any given use (e.g. analysing consumer spending requires different views of human activity than describing human task performance when driving). In the case of systems design, using a non-optimum level leads to superficial matching of design to needs if the level is too shallow, and to an inability to specify designs for needs if the level is too deep. These will be elaborated with two examples pertinent to electronic text design.

ADONIS seemed to match basic reader requirements. However, it was obvious that it did so only at a superficial level. By describing readers' needs at the gross level of 'obtaining a hardcopy', 'locating an article', 'browsing the references' and so forth it made (and matched) design targets of accurate but inadequately specified needs. The designers obviously developed a product that meets these targets, but only at the grossest level of described behaviour. A description of reading at a deeper level than this might well have produced a different set of requirements and resulted in a more usable design.

An example of a level of description too deep to specify needs for design can be found in most of the work on modelling reading by cognitive psychologists. By concentrating on theoretical structures and processes in the reader's mind or eye movements in sentence perception, word recognition and so forth, such work aims to build a body of knowledge on the mental activities of the reader. Fascinating as this is, it is difficult to translate work from this level of description or analysis to the specification of software intended to support the reading process. For a clear example of this see Rumelhart's (1977) widely acclaimed work on the development of a reading model (described earlier) and attempt to draw a set of guidelines from this that is applicable to HCI.

Highlighting limitations is only useful where it serves to advance the process of overcoming them. What is required is a description or level of discourse that bridges between these two extremes and actually supports valid descriptions of human activities in a form that is most meaningful for system design. This is not an easy task but one is helped by at least knowing where the goalposts are. Within the electronic text domain a suitable analytic framework should provide designers with a means of posing appropriate questions and deriving relevant answers. Clearly existing ones, be they psychological, typographical or information scientific, do not. How then should we conceptualize the reading process?

4.3 Identifying an appropriate abstraction

It is unlikely that the evolution of a suitable description of the reading process will result merely from performing more experiments on reading from screens. To attempt empirical testing of all conceivable reading scenarios would be impossible and, as was shown by the ADONIS evaluation, even the application of demonstrable ergonomic principles derived from such work (e.g. the importance of image quality) is insufficient to guarantee successful design.

For the purposes of designing interactive systems for process control environments, Rasmussen (1986) describes the need for a multi-layered analysis involving descriptions ranging from the social function of a system, through the information processing capabilities of the user and machine, to the physical mechanisms and anatomy of both the user and the machinery. He emphasizes the need to incorporate perspectives of human abilities from quite separate research paradigms in order to describe usefully the process of interaction with advanced technology and adds, 'it is important to identify manageable categories of human information processes at a level that is independent of the underlying psychological mechanisms' (p. 99).

In other words, the framework needed for design should not be overly concerned with the architecture of human cognition (as by definition is the case with cognitive psychological models of reading) but should address primarily the nature of the processing activity. Thus, according to Rasmussen, advances can be made on the basis of understanding the relevance of human information processing components (e.g. working memory, schemata etc.) without specifying their underlying structural form (e.g. as production systems (Anderson, 1983), blackboard architectures (Hayes-Roth, 1983) and so forth).

In HCI, the most popular interactive behaviours to examine (if judged by published studies at least) deal with text-editing, a task so heavily studied and modelled that it has earned the somewhat derogatory title in some quarters of the 'white rat of human factors'.[2] It is easy to see from such models that the Rasmussen approach of multi-layered, architecture-independent analysis is largely ignored. For example, one popular model of this activity, judged more by citations than actual use in design, is based on the cognitive complexity theory (CCT) of Kieras and Polson (1985). This theory not only formally advocates the production system architecture of human cognition as a means of 'calculating' learning difficulties in transferring between text editors, it addresses only one level of activity for the system, that of correcting spelling mistakes in previously created text. The accuracy of the model is often held up as an example to other researchers and theorists in HCI, even though its utility to designers remains, some ten years on, to be convincingly demonstrated.

Predictive modelling techniques for HCI typically rely on identifying small units of behaviour, decomposing them into their assumed cognitive primitives, analysing them with respect to time and errors, and then developing an approximate model which accounts for performance with certain boundaries such as error-free expert performance. Such models of user behaviour with technology exist not only for text editing but in less extreme forms for menu navigation (Norman and Chen, 1988), item selection with input devices (Card *et al.*, 1978), so why not reading?

The crucial point is that reading could be equivalently modelled if ergonomics were to conceptualize it as narrowly as proof-reading or item selection from a list of words à la Norman and Chen (1988). In equating reading with such activities complexity is certainly reduced but range of application is surely curtailed. Accurate models of proof-reading might eventually lead to prescriptive principles for designing screen layout and manipulation facilities for such tasks (and only such tasks), rather like the GOMS model (Card *et al.*, 1983) can theoretically aid the design of command languages for systems, but they are unlikely to prove extensible to many of the wider issues of electronic text design such as what makes a good or bad electronic text, or how should a hypertext be organized?

There is a school of thought that suggests that while such questions cannot be answered yet, the modelling approach is 'good science' and that sufficient progress in applied psychology will eventually be made by the accumulation and refinement of such low-level, predominantly mathematics-based models. Newell and Card (1985) argue that in all disciplines, hard science (i.e. technical, mathematical) drives out the soft and that quantification always overpowers qualification. With reference to HCI they argue that psychology's proper role is to use its strengths and leave behaviour outside its remit to other disciplines. For Newell and Card, the domain of psychology covers human actions within the approximate time scale of 0.1 to 10 s. Anything smaller, they claim, is covered by the laws of physics, chemistry and biology, anything larger than this but less than a matter of days, is covered by the principles of bounded rationality and the largest time frame of weeks to years is the proper subject matter of the social and organizational sciences.

Within the narrow time scale allowed psychology in their conceptualization, Newell and Card propose that psychologists concentrate on 'symbolic processing', 'cycle times', 'mental mechanics' (whatever they are!) and 'short-term/long-term memory'. They accept that the bounded rationality time-band covers many of the aspects of human behaviour relevant to HCI (and to humans in general it might be added) but 'their theoretical explanation is to be found in the interplay of the limited processing mechanisms of the psychological band and the user's intendedly rational endeavours' (p. 227).

Arguments that this form of psychology is too low level to be relevant to designers are partially correct, say the authors, but this problem will be overcome when a suitably all-embracing model of human information processing has been developed that can be applied wholesale to issues at the level of bounded rationality, i.e. the level at which human activity is normally described and understood, e.g. reading a book, writing a letter, driving a car and so forth.

This approach has been the subject of some harsh criticisms (e.g. Carroll and Campbell, 1986) and contradictory evidence. On the basis of interviewing designers of videotext systems, Buckley (1989) concluded that the type of model proposed by Card *et al.* (1983) was irrelevant to most designers. He claims that designers tend to avoid academic literature and specific experimental findings in favour of their own internalized views of typical users and good design practice. Buckley states (p. 183):

> The designers expressed concern about the ease of use of the dialogues and had some clear views of how system features under their control would affect users ... But they did not report any use of traditional forms of human factors information which are expressions of the science base and are normally represented in research papers and design handbooks.

Instead, he found that designers in his research relied heavily on 'pre-existing internalized frameworks' (p. 184) which consist of primitive and weakly articulated models of users and their tasks that the system they were designing was intended to support. Buckley (1989) goes on to emphasize the importance of providing information to designers in a form compatible with this style of working. Such findings are not unique, similarly doubtful views of the validity of formal models and standard human factors literature based on empirical findings have been expressed by other researchers who have interviewed designers over the last decade (e.g. Hammond *et al.*, 1983; Eason *et al.*, 1986; Dillon *et al.*, 1993a).

Eason *et al.* (1986) report data from a survey of designers which examined the use they had made or intended to make of the various outputs from human factors research. In accordance with Buckley's data, they report low-level awareness and use of academic literature such as journal papers or experimental data and zero usage or willingness to use formal models of the kind advocated by Newell and Card (1985). However, the designers they surveyed were willing to receive training on user-centred design methods and read appropriate guidelines. Dillon *et al.* (1993) report similarly low usage of formal modelling techniques and no request for improved ones. Instead, the designers they surveyed across Europe still sought further guidelines, usability standards and goals that could be tested for, and where also prepared to undergo some training in user-centred design methods.

Such findings partly confirm the conclusions drawn from the ADONIS study where it was obvious that designers had some ideas of the users they were designing for, except that in this case, they were obviously also aware of some of the recommendations from the literature. This contrasts sharply with Buckley's designers, some of whom seemed surprised when he told them such a literature actually existed. Regardless of their familiarity with the literature though, designers must have an idea of who they are designing for and what tasks the system will support, how else could they proceed? Their views are naturally partial and often intuitive. Therefore making this conceptualization more explicit and psychologically more valid in an appropriate way would seem to be of potential benefit to the software interface design world.

The second weakness in Newell and Card's argument is that it assumes the design world can afford to wait for an all-embracing cognitive model to emerge while all around us, technological advancement accelerates. They counter this criticism with the somewhat surprising statement that technology does not advance as fast as many think it does. Though they refer to paradigm interfaces, such as the standard graphical user interface found on many systems, as evidence that stability can be seen in design, it is less easy to see how their conceptualization of HCI can either take us beyond current perspectives on usability or help ensure innovative applications are acceptably designed.

Regardless of the level of advancement, in the domain of reading at least, cognitive psychological models of the process exist which satisfy many of the criteria of hard, quantitative science (e.g. Just and Carpenter, 1980) but as has been repeatedly pointed out, these just do not seem to afford much in the way of design guidance now. What seems to be required is a descriptive level above the information processing models advocated by Newell and Card, but below the very high level descriptions of the bounded rationality approach favoured by information scientists. This is one of the levels we could expect to find in adopting Rasmussen's architecture-independent approach. In the case of reading and electronic text systems a suitably embracing framework would need to cover the range of issues from why an individual reads to how the screen can be best laid out, which would naturally induce inputs from a variety of research paradigms. However, these inputs would need to be organized and conceptually clarified in a manner suitable for designers.

4.4. Conclusions and the way forward

It has been argued in this chapter that many of the problems inherent in electronic text design spring from the lack of a design-relevant description of the reading process. Cognitive psychology in the main, but information science as well, have been criticized for providing unsuitable levels of abstraction at which to describe the human behaviour relevant to design. This is, however, less of a criticism of either discipline which do not exist to serve the needs of software designers, but more an indictment of human factors researchers' failure to provide their own theories. Barber (1988) remarks that ergonomics as a discipline has relied so heavily on theories borrowed from other disciplines, that members of the human factors community see no need to develop their own and remain content to draw on standard psychological or physiological theories as required. If one could be guaranteed that standard theories could be drawn on this way and were never distorted by such application, this position might be tenable but such guarantees cannot be provided by any theory.

The practical question then is what would a human factors practitioner have added, for example, to the ADONIS design to make it more usable had she been involved as early as the specification stage? The simple truth of the matter is that deriving a more specific set of ergonomic criteria from the literature would have been difficult. The specification clearly included reference to most obvious variables. What would have been required to improve ADONIS is a user-centred design process of the form outlined in Chapter 2, involving stakeholder and user analysis, the setting of usability targets, iteration through prototypes, and continuous evaluations until the design targets were met.

The problem with this approach is that it is costly in terms of time and resources. What needs to be included is some means of constraining the number of iterations required. This is best achieved by ensuring that the first prototype is as close to the target as possible. Of necessity, this would have involved carrying out user requirements capture and task analyses of readers interacting with journals and searching for articles. Output from such work would have been fed back to the designers to guide decisions about how the prototype interface should be built. Subsequent evaluation would then have refined this to an even better form.

It is almost certain that such work would have led to a better design than the current one, from which we can conclude that the type of knowledge generated by task analyses and prototype evaluations is directly relevant to design. The questions then become, what form of knowledge is this, at what level is it pitched, can it be extracted earlier and, more importantly, can a generalized form be derived to cover all reading situations?

An attempt to provide answers to questions of this type is made in this book. The primary means of providing them will be to examine the inputs made by the author to the design of real-world document systems developed or tested at HUSAT as part of a variety of funded projects. By using this work as a background it is possible to identify the type of human factors inputs needed and found to be useful in real design projects.

If the results from any one document design project are to be generalizable to others however, it is important to know how work on one text type differs from or is similar to others. Without such knowledge it would not be possible to make any meaningful generalizations about electronic text design from any one study or series of studies on a text. Unfortunately, there is as yet no agreed classification scheme for describing the similarities and differences between texts. To overcome this, a suitable classification scheme must be developed as a first stage of the work in deriving a framework for designing electronic texts. This is in line with other views. As de Beaugrande (1981, p. 297) puts it, 'To adequately explore reading, a necessary first step is a firm definition of the notion of "text" – it is not just a series of sentences as one is often required to assume'. He goes on to say, 'It follows that reading models will have to find control points in the reading process where text-type priorities can be inserted and respected' (p. 309).

To this end, the question of text type is addressed first and an investigation into readers' own classification systems of the world of texts is reported in the following chapter. This will be used to provide a basis to subsequent work and offer a means of generalizing beyond the particulars of any one particular text type.

Notes

1. The survey of citation style revealed that users tend to refer to articles in the form author/year/title/journal, or author/journal/article/year. ADONIS structured input in the form: ADONIS number/ISSN number/journal/year/author etc. which was considered very confusing by some users and led to frequent errors during trial tasks (see Dillon, 1988).
2. This descriptor was first seen by the author in an item on HICOM, the electronic conferencing system of the British human factors community. The implication of the description is surely that text-editing tells us as much about human–computer

interaction as a rat's performance tells us about being human. Depending on theoretical perspective that might mean a lot or a little, but given the absence of hardline behaviourists in HCI research one can only conclude that the writer intended it as a little.

5

Classifying information into types:
the context of use

All good books are alike in that they are truer than if they had happened.
Ernest Hemmingway, Old Newsman Writes, *Esquire* (1934)

5.1 Introduction

Classification of concepts, objects or events is the hallmark of developed knowledge or scientific practice and to a very real extent, typologies can be seen as a measure of agreement (and by extension, progress) in a discipline. Fleishman and Quaintance (1984) point out that in psychology, a real problem in applied work is transferring findings from the laboratory to the field and that the lack of appropriate typologies (of tasks, acts, events and concepts) is a major stumbling block. As demonstrated in the literature review in Chapter 3, when it comes to studies of reading there is no standard reading text or task that can be used to investigate all important variables and thus, despite heavy reliance on proof-reading short sections of prose (and there exists wide variability even in this type of task) we are not in a very good position to generalize effectively from experimental to applied settings.

de Beaugrande's (1981) call for an understanding of text types has been echoed often in the field of electronic text and hypermedia where it is felt by some that a useful typology of texts could aid distinctions between potentially suitable and unsuitable electronic texts (Brown, 1988; McKnight *et al.*, 1989). Such a typology would presumably ultimately provide a practical basis for distinguishing between the uses to which different texts are put and thus suggest the interface style required to support their hypertext instantiations.

There are a number of distinguishing criteria available to any would-be typologist such as fiction and non-fiction, technical and non-technical, serious and humorous, narrative and expository or like Hemmingway, good and bad. These may discriminate between texts in a relatively unambiguous manner for some people and the list of such possible dichotomies is probably endless. Such discriminations however are not particularly informative in design terms. In this domain it is hoped, and even expected, that a typology could distinguish texts according to some usage criteria, leading to a situation where evidence can accumulate to the point where the designer could map interface features to information types reliably. In such a scenario, for example when designing any electronic document, it would be possible to select or reject certain design options that the typology suggests are appropriate or inappropriate for that text type. This contrasts sharply with current knowledge where any attempt to identify features known to have worked with say, SuperBook, might have no relevance to someone designing an electronic catalogue. Of course, as pointed out in Chapter 2, it is inappropriate to assume usability can ever be so prescriptively linked to interface variables in some form of inheritance mapping in a text or task scheme but the point here is that ideally, a relevant typology should group information according to aspects of the reader-text interaction that support valid generalization.

de Beaugrande (1980, p. 197) defines a text type as 'a distinctive configuration of relational dominances obtaining between or among elements of the surface text, the textual world, stored knowledge patterns and a situation of occurrence'.

This is a rather complex definition invoking the relationship between how the text looks, the readers' knowledge and experience as well as pointing to the context in which the text is met. The term 'textual world' is interesting and points to the subject matter of the document as well as its relationship to other documents. It is not clear though, how relational dominances of this kind can be assessed. By way of illustration of this view of text types, de Beaugrande (1981, p. 278) offers the following examples: descriptive, narrative, argumentative, literary, poetic, scientific, didactic and conversational. However, de Beaugrande freely admits that these categories are not mutually exclusive and are not distinguishable on any one dimension. As he says, 'many text types are vague, intuitive heuristics used by readers to tailor their processing to individual examples'.

This is interesting as it raises the possible distinction between text types as objective text-based attributes and as subjective reader-based perceptions. That is, can a typology be based on (more or less) objective criteria determinable by an examination of the published artefact or can one only be drawn up by taking account of readers' perceptions of type? And if the former, is it possible that constituent elements could be identified that supported the future classification of emerging electronic forms? It would be convenient if the former was the case but it is more likely that, as in all things concerning humans, real insight is only gained by assuming the latter, i.e. we will need an understanding of the readers' perceptions.

Typographers have a long-standing professional interest in text types. By and large the discipline seems to categorize information on objective criteria, i.e. typical typographic classifications relate to demonstrable attributes of the physical text. Waller (1987) for example proposes analysing text types in terms of three kinds of underlying structure:

- topic structure, the typographic effects which display information about the author's argument, e.g. headings;

- artefact structure, the features determined by the physical nature of the document, e.g. page size;
- access structure, features that serve to make the document usable, e.g. lists of contents.

According to this approach, any printed artefact manifests a particular combination of these three underlying structures to give rise to a genre of texts. This emergent fourth property is termed the conventional structure. As the name suggests these underlying structures give rise to the 'typical' text we understand as 'timetables', 'newspapers', 'novels' or 'academic journals'.

Like de Beaugrande, Waller describes the three structural components of his categorization as heuristic concepts. However, for him this appears to have less to do with the reader approaching the artefact (although this is an element in Waller's categorization) than with the means of providing a basis for genre description. From an ergonomic perspective it is concerned less with the readers and their conceptualization of the text therefore than with layout, presentation and writing style. Thus a mapping of this analysis to the design of electronic texts is not logically determined even though the categories of structure outlined in this approach seem immediately more relevant to ergonomic interests than de Beaugrande's and the concept of access structure has certainly been embraced in discussions of electronic text design (e.g. Duffy *et al.*, 1992).

Inter-disciplinary boundaries in this domain are not always clear or even desirable. Cognitive psychologists have taken increasing interest in the relationship between so called 'typographical' features and the reading process (e.g. Hartley 1985) and typographers look to psychology for theoretical explanations of typographic effects on humans. This inter-disciplinarity has not always been so positive however. Waller (1987) reports on a long-standing debate in the typographic discipline on the value of applied psychological research (largely the univariate, experimental model manipulating text attributes such as line length or font). Criticisms of this type of work, typified early on by people such as Tinker (1958), have existed since the 1930s (e.g. Buckingham 1931) and in nature and content largely resemble current arguments in the field of HCI where the value of human factors experimental research to system designers is often questioned.

Ergonomics invariably attempts to draw heavily on psychological theories in extending its reference beyond the immediate application and it is clear that some work on text groups exists in contemporary psychology. In psycho-linguistics for example, van Dijk and Kintsch (1983) use the term 'discourse types' to describe the superstructural regularities present in real-world texts such as crime stories or psychological research reports. Their theory of discourse comprehension suggests that such 'types' facilitate readers' predictions about the likely episodes or events in a text and thus support accurate macro-proposition formations. In other words the reader can utilize this awareness of the text's typical form or contents to aid comprehension of the material. In their view, such types are the literary equivalent of scripts or frames and play an important role in their model of discourse comprehension.

While they do not themselves provide a classification or typology, this work is directly relevant to the design of electronic texts as will be discussed in detail in a later chapter.

Wright (1980) described texts in terms of their application domains and grouped typical documents into three categories: domestic (e.g. instructions for using appliances), functional (e.g. work-related manuals) and advanced literacy (e.g. magazines or novels). She used these categories to emphasize the range of texts that exist and to highlight the fact that reading research ought to become aware of this tremendous diversity. This is an important point, and one central to ergonomic practice. Research into the presentation and reading of one text may have little or no relevance to, and may even require separate theoretical and methodological standpoints from, other texts. Consider for example, the range of issues to be addressed in designing an electronic telephone directory for switchboard operators with no discretion in their working practices, and an electronic storybook or 'courseware' for children aged 8–10. Rigid implications on interface design drawn from one will not easily transfer to the other, unless one at least possesses the appropriate perspective for understanding the important similarities and differences between them. A valid typology of texts should be useful in such a situation as researchers from a range of disciplines have suggested.

5.2 Reader-perceived distinctions between texts

What all the classifications outlined above failed to take account of however, is the readers' views of these issues and it is with this concern in mind that a colleague and I decided to carry out an investigation of readers' perceptions of text types (Dillon and McKnight, 1990). This marked a first attempt at understanding types purely according to reader-perceived differences. In so doing it was hoped that any emerging classification criteria would provide clues as to how electronic documents can best be designed to suit readers, an intention not directly attributable to any of the aforementioned categorizations.

It was not immediately obvious how any researcher interested in text classification schemes should proceed on this matter however. Few categorizers to date have made explicit the manner in which they derived their classifications. It seems that regardless of theoretical background, most, if not all, have based their schemes on their own interpretations of the range of texts in existence. True, their classifications often seem plausible and the knowledge and expertise of all the above-cited proposers is extensive; yet it is impossible to justify such an approach in the context of system design. It could be argued that readers do perceive the manifestation of the different structures or heuristics even if they cannot reliably articulate them. After all, the conventional genres are real world artefacts. However, this can only be assumed, not proven as yet. This section reviews the Dillon and McKnight work and in so doing reports some readers' perceptions of text types.

In the first instance it must be said that we soon realized that recognizing the need for a more objective means of classification was one thing, identifying a suitable technique which enables this was another. To derive a reader-relevant classification necessarily requires some means of measuring or scoring readers' views. The available methodological options are techniques such as interviewing, questionnaires or developing some form of sorting task for readers to perform on a selection of texts. The case for using a questionnaire is flawed by the absence of any psychometrically valid questionnaire on this subject and the effort involved in developing one within the time-scales of any one investigation.[1]

Interviewing individuals is a sure method of gaining large amounts of data. However making sense of the data can prove difficult both in terms of extracting sense and in overcoming subjective bias on the part of the interviewer. While the latter problem can be lessened by using two skilled interviewers rather than one and careful design of the interview schedule, the former problem is more difficult to guard against in this context. Text classification as a subject, could be sufficiently abstract as to cause interviewees problems in clearly articulating their ideas in a manner that would support useful interpretation of the data. With these issues in mind we decided to consider some form of sorting task.

Repertory grid analysis was eventually chosen as the most suitable technique of this type for eliciting suitable data. Developed by George Kelly (1955) as a way of identifying how individuals construe elements of their social world, Personal Construct Theory (PCT) assumes that humans are basically 'scientists' who cognitively (or more accurately, in Kelly's terms, 'mentally') represent the world and formulate and test hypotheses about the nature of reality (Bannister and Fransella, 1971).

The repertory grid technique has been used for a variety of clinical and non-clinical applications (e.g. studying neurotics: Ryle, 1976; job analysis: Hassard, 1988) and has been applied to the domain of Human Computer Interaction, particularly with respect to elicitation of knowledge in the development of expert systems (Shaw and Gaines, 1987). Its use as an analytic tool does not require acceptance of the model of man which Kelly proposed (Slater, 1976). However, the terms Kelly used have become standard. Therefore one may describe the technique as consisting of 'elements' (a set of 'observations' from a universe of discourse), which are rated according to certain criteria termed 'constructs'. The elements and/or the constructs may be elicited from the subject or provided by the experimenter depending on the purpose of the investigation. Regardless of the method, the basic output is a grid in the form of n rows and m columns, which record a subject's ratings, usually on a 5- or 7-point scale, of m elements in terms of n constructs.

The typical elicitation procedure involves presenting a subject with a subset of elements and asking them to generate a construct which would meaningfully (for them) facilitate comparison and discrimination between these elements. The aim is to elicit a bipolar dimension which the subject utilizes to comprehend the elements. In the present context, all participants were given a similar set of texts (the elements) which they used to generate distinguishing criteria (the constructs). As constructs are elicited and all elements subsequently rated on these, a picture of the subject's views and interpretations of the world of documents emerged. What follows is a summary of the results published by Dillon and McKnight (1990).

5.3 Readers, texts and tasks

Six professional researchers had grids elicited individually. Elements, which were identical for all consisted, of nine texts:

- a newspaper (*The Independent*, a UK 'quality' daily broadsheet)
- a manual (MacWrite Users Guide)
- a textbook (*Designing the User Interface:* Ben Shneiderman)
- a novel (*Steppenwolf*: Herman Hesse)
- a journal (*Behaviour and Information Technology*)
- a catalogue (Argos Catalogue Spring 1988 – personal and household goods)
- a conference proceedings (*CHI '88*)
- a magazine (*The Observer Sunday Colour Supplement*), and
- a report (HUSAT Technical Memo)

Constructs were elicited using the minimal context form (Bannister and Mair, 1968) which involves presenting people with three elements, known as the triad, and asking them to think of a way in which two of these are similar and thereby different from the third. When a meaningful construct was generated the two poles were written on cards and placed either side of a 1–5 rating scale on a desk. Researchers then rated all the texts according to the construct, physically placing the texts in piles or individually on this scale.

5.4 Results

Full details of the analysis can be found in Dillon and McKnight (1990). It is sufficient here to describe the resulting grids from the FOCUS program used (Shaw, 1980). A focused grid for one subject is presented in Figure 5.1. The grid consists of the raw ratings made by the subjects with the element list above and the construct list below. The FOCUS program automatically reorders these to give the minimum total distance between contiguous element and construct rating columns. Dendrograms are constructed by joining elements and constructs at their appropriate matching levels.

In Figure 5.1, the elements are on top and the constructs are to the right of the reordered ratings. The matching levels for both are shown on adjacent scales. High matches indicate that the relevant elements share identical or similar ratings on the majority of constructs or the relevant constructs discriminate identically or similarly between the majority of elements. Thus in Figure 5.1 it can be observed that the journal and conference proceedings

Designing usable electronic text

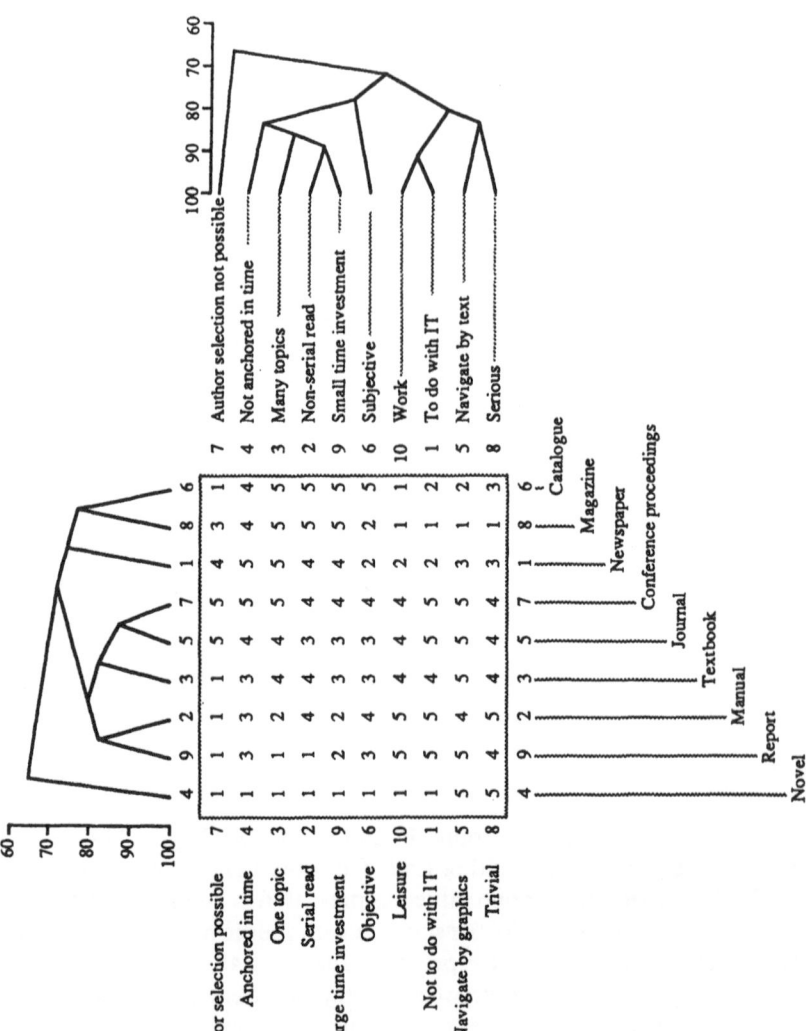

Figure 5.1   A FOCUSed grid for one reader.

match highly and that the novel is least similar to the others. Criteria such as 'Work – Leisure' and 'To do with IT – Not to do with IT' offer the highest match among the constructs elicited, while 'Single author selection possible – not possible' is the lowest match. In other words the journal and the conference proceedings are seen as very similar to each other but very different from the novel by this reader, and every time a text is described as work related, it also tends to be described as being about information technology. By proceeding in this manner it becomes possible to build up a detailed picture of how an individual construes texts. In the present study all six grids were analysed (focused) together as one large grid. The focused collective grid are presented separately in the element and construct trees in Figures 5.2 and 5.3.

5.4.1 The texts

The texts (elements) clustered into three distinct groups (Figure 5.2). These were the work related, the 'news' type texts and the novel. The highest match was between the conference proceedings and the journal (90.2 per cent) followed by the newspaper and the magazine (85.1 per cent). Basically this means that any time, for example, the journal was rated as being high or low on construct X then the conference proceedings were rated similarly. The textbook and report both joined the first cluster at more than the 82 per cent matching level. This cluster eventually incorporated the software manual at the 62 per cent level suggesting that while this manual shared some of the ratings of the other elements in that cluster it was noticeably different from them on certain constructs.

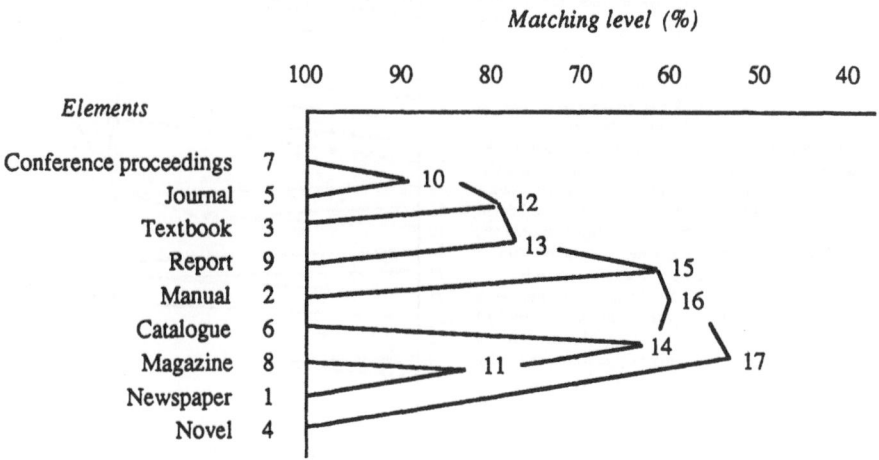

Figure 5.2 Dendrogram of element clusters for all readers.

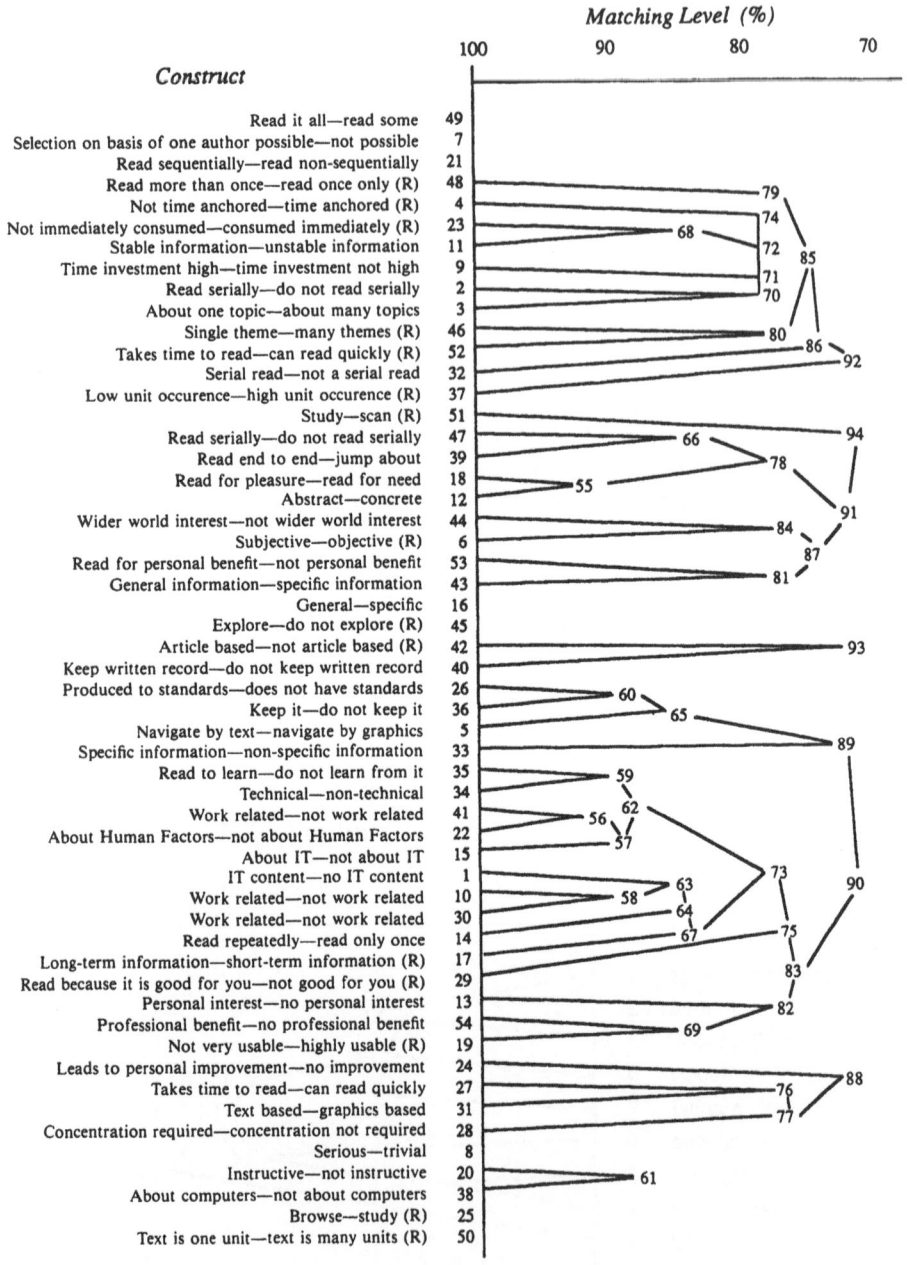

Figure 5.3   Construct clusters for all readers.

The catalogue matched the newspaper and magazine at 69.4 per cent suggesting that it is perceived as similar in many ways to those types of text. The novel however was the last element to be incorporated in a cluster, only linking with other elements at the 53.2 per cent level, by which time all the other elements had formed one large cluster. This suggests that it was perceived as a unique text type among all these elements.

5.4.2 The readers' distinguishing criteria

Fifty-four constructs were elicited from this sample. In order to ensure that only tight clusters were identified a minimum matching level of 70 per cent was defined as the criterion. This is somewhat arbitrary and at the discretion of the investigators, the lower the criterion the more matches are produced until eventually all constructs would match at some point. The construct dendrogram is presented in Figure 5.3. Dillon and McKnight (1990) identified three major construct clusters from this analysis which are outlined below.

Cluster 1: Work related material

This cluster described texts which are work related, about human factors or IT, contain technical or specific information and would be read for learning or professional purposes.

Every researcher distinguished between work related and personal reading material. All of their constructs about this distinction matched up at the 77.7 per cent level. Obviously construing texts in terms of their subject matter and relevance to work was common to all these researchers. The constructs 'reading to learn' and 'technical' matched at 88.8 per cent and joined up with 'work related'/'about human factors'/'IT related' at that level too. Also contained in this cluster were 'read repeatedly' and 'long-term information', matching a work related sub-cluster at the 83.3 per cent level. An element that was prototypical of this construct cluster was the journal. A very poor match with this cluster was observed for the newspaper and the magazine.

Cluster 2: Personal reading material

This contained texts that were seen as personal reading material, containing general or abstract information that would be read in a serial fashion.

The highest match in this cluster was between the constructs 'abstract – applied' and 'reading for pleasure – reading for need' which matched at the 94.4 per cent level. The next highest was at the 83.3 per cent level between 'serial' and 'read from end to end'. These pairs of constructs then joined at the 77.7 per cent level. The constructs 'personal benefit', 'general information' 'subjective' and 'wider-world interest' all matched at the 77.7 per cent level. These sub-clusters all joined at the 72.2 per cent level.

These constructs suggest that certain texts are seen as more personal than work related and contain information that is general in nature or subjectively interesting. The presence of constructs that indicate they are read in a serial fashion would indicate texts that are not intended for reference but for complete reading. A text that closely matched most of these descriptors was the novel. A very poor match with these constructs was the catalogue.

Cluster 3: Detailed lengthy reading material

This cluster described texts that were seen as having one main subject or topic, the content of which is stable and requires a high investment of time to read. Such texts are also characterized by serial reading.

The highest match in this cluster was between the constructs 'not immediately consumable' and 'contains stable information' which matched at the 83.3 per cent level. The constructs 'read serially' and 'one topic' matched at the 77.7 per cent level, as did 'time investment to read is high' and 'single theme'. The constructs 'read more than once' and 'not anchored in time' matched at 77.7 per cent and all these constructs were joined by another construct 'serial' at this level too. The final construct in this cluster above the criterion level was 'low occurrence of separate units in the text' which joined all of the other constructs at 72.2 per cent level. A text that closely matched these constructs was the novel; the newspaper and magazine were typically rated as the opposite of these.

The results demonstrate that people's manner of construing texts is complex and influenced by numerous factors. Clear distinctions between texts such as 'fiction and non-fiction' are too simplistic and superficial. On a psychological level individuals are more likely to make distinctions in terms of the type of reading strategy that they employ with a text, its relevance to their work or the amount/type of information that a text contains. This is not a basis for classification used by many proponents of text typologies.

While the terms or descriptors employed and their similarities or differences (the face validity of the output) are interesting, it is their treatment of the elements that is ultimately important. The elements: textbook, journal, conference proceedings and report were all matched highly forming a particular cluster of text types. The magazine and newspaper were also matched highly. These are reasonable groupings between what may broadly be termed 'work' and 'leisure' texts. The novel is the one text type that matches least well with all the others and once again, this appears sensible. Examining the constructs that distinguish between these texts can shed more light on the classification criteria employed by these readers.

The journal and textbook types were described unsurprisingly as work-related, being about human factors or IT, containing specific or technical information and were read for professional benefit or in order to extract specific information. They were likely to be read more than once and be of long-term rather than immediate use or relevance. This distinguishes them from the other two element clusters which were more likely to described as read for leisure and containing general, subjective, or non-technical information. The novel was further distinguished from the newspaper and magazine by the need to invest a lot of time and read it serially and completely.

On the basis of these results Dillon and McKnight (1990) argued that texts seem distinguishable on three levels which any one construct may reflect:

- **why** they are read – for professional or personal reasons, to learn or not, out of interest or out of need, etc.;
- **what** type of information they contain – technical or non-technical, about human factors or not, general or specific, textual or graphical, etc.;
- **how** they are read – serially or non-serially, once or repeatedly, browsed or studied in depth, etc.

By viewing text types according to the various attributes of these three levels of use it is easy to distinguish between e.g. a novel and a journal. The former is likely to be used for leisure (**why**), contain general or non-technical information (**what**) and be read serially (**how**), whereas the latter is more likely to be used for professional reasons (**why**), contain technical information which includes graphics (**what**) and be studied or read more than once (**how**).

Importantly, there is an individualistic aspect here. The same text may be classified differently by any two readers. Thus a literary critic is likely to classify novels differently from a casual reader. Both might share similar views of **how** it is to be read – serially or in-depth – but differ in their perceptions of **why** it is read or **what** information it contains. The critic will see the novel as related to work while the casual reader is more likely to classify it as a leisure text. What it contains will differ according to the analytic skill of the reader with a critic viewing e.g. Joyce's *Ulysses* as an attempt to undermine contemporary English and the casual reader (if such exists) seeing it as a powerful stream of consciousness modern work. Neither is wrong; in fact both are correct. Any classification of texts based on psychological criteria must, by definition, allow for such individual differences in perception and meaning.

Readers may vary their classification of texts according to tasks (i.e. within-subject differences). Several researchers remarked that some texts could possibly be classed as work-related and personal reading depending on the situation. An obvious example of this occurs when someone reads an academic article that is both relevant to one's work and intrinsically interesting in its own right. For individuals whose professional and personal interests overlap such an occurrence was common. The present categorization of texts allows for this by placing emphasis on the motivation for reading (the **why** axis).

Allowing for both between- and within-subject variance reflects the underlying psychological nature of the categorization. A classification based solely on demonstrable objective distinctions is likely to have either a very limited sphere of application outside of which it loses relevance or be as simplistic as distinctions of the form paperback – hardback or fiction – non-fiction, etc. The present classification does not therefore provide a genuine typology of texts, but it does offer a way of distinguishing between them and analysing readers' perceptions. It suggests that three aspects of the text are important to the reader in categorizing a document or published artefact.

5.5 Implications for electronic text design

How does all this relate to the development of electronic text systems? Theoretically, at least, one could seek to determine the complete membership of each attribute set by identifying all possible **hows**, **whys** and **whats** through some form of task analysis and subsequently plot the range of texts that match any combination of the three. However, such a level of analysis is probably too fine-grained to be worth pursuing. The perspective proposed here is best seen as a simple representation of the factors influencing readers' perceptions of texts. Characterizing text according to **how**, **why** and **what** variables can provide a means of understanding the manner in which a given readership is likely to respond to a text. So, for example, it could be used for understanding the similarities and differences between a telephone directory,

> **why:** to contact a friend
> **what:** specific numeric data
> **how:** skim and locate

and a novel,

> **why:** leisure
> **what:** large text, no graphics
> **how:** serial, detailed read

and thus quickly facilitate sensible decisions about how they should be presented electronically. Given what we know about reading from screens and HCI, the position of a novel in such a classification system would suggest that an electronic version would not be used frequently whereas an electronic telephone directory may usefully be designed to ease the process of number location or aid searching with partial information.

At a gross level, such a classification may serve to guide decisions about the feasibility of developing a usable electronic version of a text type. Where the likely readership is known this can act as a stimulus to meaningful task analysis to identify how best to design such texts. In the example of the novel above, task analysis might reveal that the novel is required for teaching purposes, where sections need to be retrieved quickly and compared linguistically with other sections or writers. Obviously this would alter the **how** and **why** attributes of the previous classification, indicating that an electronic version is now more desirable, e.g.

> **why:** education in literary studies/text analysis
> **what:** large text, no graphics
> **how:** location, short section reading and comparison

This highlights how such a categorization could lead to a more informed analysis of the design potential than deceptive first impressions and common sense suggest. Hypertext novels could in fact be a boon to teachers in a variety of educational scenarios yet even expert human factors professionals might not see that at first glance. Norman (1988, p. 235) for example, when discussing the potential for hypertext applications in general said: '[Hypertext] is a really exciting concept. But I don't believe it can work for most material. For an encyclopaedia, yes: or a dictionary; or an instruction manual. But not for a text or a novel.'

Regardless of what he means by 'a text' in this quotation, it is clear that a more critical examination of the **why** question, the motivation or reason for reading (in this case novels) indicates that a hypertext version could work. Obviously even human factors experts cannot be relied on to see the human implications of any technology in advance.

By knowing more about why individuals access texts, how they use them in terms of reading strategies and the distinctions they make between the information type presented, it should prove possible to be more specific about the question of text types. Traditionally the **how** question has been the domain of the psychologist, the **what** question the domain of the typographer and the **why** question has been largely ignored. The results from this study strongly suggest that this is a mistake.

Texts possess more than purely physical properties. Readers imbue them with values and attribute them roles in the support of a host of real life acts in innumerable work and leisure domains. This is not immediately obvious from a physical examination of a text and indeed cannot be understood by any reader immediately. Rather these are acquired skills, perceptions shaped by years of reading experiences and the responses of authors and publishers in attempting to provide support for reading.

Reading is purposeful and intentional. Texts can support many such interpretations for the reader depending on context. As I have attempted to show in the present chapter, what is a relaxing serial read for one reader may be a jumping quick scan for another. As such, it is nonsense to assume we can devise prescriptive rules for designing standardized electronic versions of certain text types such as novels, journals or catalogues (or other conventional genres). One should perhaps realize that one cannot typify texts for readers, only generalize reading contexts with their constituent interactive components of reader, task, text and environment. Thus the genres we require for electronic texts are ones embodying criteria of usage such that they are grouped according to the tasks they support, the reasons for the interaction and the nature of their contents. In other words there is a coupling between a text and the reader and this may shift as the reader's task alters. Reliance solely on existing convential genres is misleading since any one genre may exist in multiple contexts of use.

It is conceivable that utilizing this perspective on information types could support the design of new text types. In other words, by conceptualizing the reader-text context of use appropriately according to **why**, **what** and **how** criteria, documents could be created that support the reader's interaction with the information. This would be a natural extension of user-centred design methods – creating documents specific to user needs rather than seeking always to modify existing types. However, this is unlikely to be so simple since, as I will attempt to show later, existing genres are potent formats that in themselves impose and support much by way of use.

5.6 Conclusions and the way forward

The variance in texts that readers regularly utilize was identified as a natural starting point for any analysis of the real-world reading process, and an area requiring specific attention in the context of electronic text systems. The present chapter reviewed several approaches to describing texts in a manner suitable for discussing the design of electronic versions. To this end, a description based on three reader-perceived characteristics seems most appropriate: the **why, what** and **how** aspects.

These aspects represent readers' own classification criteria and also offer a means of describing texts in a way that is directly related to designing electronic versions. If a book is accurately described according to these criteria, it should lead to specific issues for consideration in design in a way that would not necessarily be the case for a description based on more traditional criteria. Specifically, by focusing attention on how a text is read, immediately leads to a consideration of task issues; by examining what a text contains the questions of content and structure are addressed; and by focusing on the **why** aspect, the context and motivation of reading are highlighted. Furthermore, by appreciating the differences in texts in these terms one is in a better position to judge the likely relevance of experimental findings on one text to another. The real test of this approach however, is the extent to which meaningful data can be derived from such a classification. In the following section, empirical applications of this approach are described.

Notes

1. A questionnaire developed according to sound psychometric principles is a lengthy task involving stages of item generation, selection, piloting and analysis, possibly through several iterations before a reliable and valid tool is developed (Oppenheim, 1966). This distinguishes questionnaires from the more loosely created 'questions on a page' type of survey common to human factors or market research.

6

Capturing process data on reading – just what are readers doing?

Child, do not throw this book about,
Refrain from the unholy pleasure
Of cutting all the pictures out.
Preserve it as your chiefest treasure.
Hilaire Beloc, Bad Child's Book of Beasts (1896)

6.1 Introduction

The reader-elicited text descriptors outlined in the last chapter provide a starting point for a consideration of the design issues for electronic text. They reflect both cognitive and behavioural aspects of the reader–text interaction rather than more common classifications based on publication genre or subject matter. Focusing attention on why and how texts are read as well as the reader's views of what texts contain, immediately brings forth issues related to task, motivation for reading, readers' models of the information space and the reading situation; factors certain to be of importance in the ultimate success or failure of any presentation medium.

It is a simple enough matter to describe a text according to these three criteria if the description consists only of the type of one-liner provided in the examples of the last chapter. However, such descriptions are not enough to provide a basis for specifying software where details of a more precise nature are required (e.g. Easteal and Davies, 1989). Furthermore, merely describing texts in this way as a result of introspection or best guess on the part of the designer or human factors expert is far from optimum, though such uses might be appropriate for the initial consideration of issues in a design prior to formal specification. It is clear that designers are normally not the best source of information on how typical users of their products view the situation. What is required therefore is the demonstration that this approach can be utilized to gather evidence of reader behaviour (particularly process data) and that the resulting output has relevance to system design.

6.2 The range of texts to describe

In the course of design work on electronic documents at HUSAT to date, four distinct document types have been subjected to this form of descriptive analysis: academic journals, software manuals, research project material (a collective description of the reports, data and documented communications of a team of researchers on one project) and a handbook of procedures for a multinational manufacturing company. This chapter will concentrate on the academic journals and software manuals. The research project material is covered in McKnight *et al*. (1990b) while the work on the industrial procedures handbook remains confidential.

Much of the work reported in this book was carried out in parallel with, and often as part of, a series of research and development projects designing and evaluating electronic documents at HUSAT. The investigation into journal usage described here, had direct relevance to real design issues for the team designing an electronic journal application and in fact the investigation was driven by the need for information of this kind. Software manuals were selected for analysis as it was felt that they are a frequently used text for which electronic versions partly exist (in the form of on-line help facilities) and are likely to be increasingly required electronically. Furthermore they satisfy the criterion of distinction from journals as a text type according to the results of the repertory grid study thereby offering a useful indication of the breadth of application for the **why, what** and **how** approach.

6.3 Presenting journals and manuals electronically: a brief history

The idea of presenting academic journals in electronic form is not new. Indeed such journals have been empirically investigated at HUSAT since the early 1980s and the first electronic journal project (the EIES project) was established in America as far back as 1976 (Sheridan *et al.*, 1981). The British Library-funded BLEND (Birmingham and Loughborough Electronic Network Development) project (Shackel, 1990) ran from 1980 to 1984 and examined the potential for information technology to support an electronic journal that not only presented the text on screen but also facilitated submission and refereeing of papers electronically. That project highlighted the need for improved screen technology and text manipulation facilities independently of the empirical literature that was only beginning to emerge at the time (Pullinger, 1984; Shackel, 1987).

According to Meadows (1974), among the aims of academic journals are: aiding the flow of information and establishing priority of ideas and findings; therefore electronic versions may have a significant part to play (particularly given the long publication times of many journals). Other advantages of electronic over paper journals lie in their ease of storage, rapid access and convenience of searching. Libraries across the world need to create miles of new shelf space annually merely to house new publications, and the low usage statistics for many academic journals suggest that for a substantial amount of time, academic journals are sitting idle. A recent review of libraries in the Times Higher Education Supplement in the UK reported that a square foot of shelf space in central London costs £75 ($120 or so) per annum for academic journals that will likely be accessed less than six times a year. Yet popular articles are often in much demand and the absence of a single copy of a journal from a shelf (because say, someone else has taken it to read) can cause immense frustration to busy scholars, scientists and students. Electronic versions stored on CD-ROM and accessed at the point of a mouse and push of a button, offer the promise of unlimited and immediate availability to the reader. However, as with all technological promises, potential disadvantages stem from the inherent ergonomic problems associated with reading from screens as well as issues of copyright.[1]

Software manuals accompany many, if not all off-the-shelf or bespoke applications and it is a part of the folklore of human factors that they are invariably overlooked by most users. Carroll (1984) reviews some evidence of this phenomenon and while it is now a cliché to state that 'manuals are never read,' there is little evidence to support that jaundiced view, even if it does spring from the general recognition of lower use than most technical authors would care to admit. On-line help facilities are becoming increasingly sophisticated and where it was once suggested dismissively that most help facilities were merely electronic versions of the paper manual, the development of query-in-depth (QID), context-specific presentations with hypertext qualities has now created a more favourable view of the concept of electronic manuals.

Electronic manuals need not just accompany software. Manuals for hardware and non-computer based machinery, either repair or operation instructions, offer potential application domains for electronic text. When one considers estimates that nuclear submarines or spacecraft carry more weight in, or provide more space to paper documentation than they do any other item (e.g. Ventura, 1988), then the potential for electronic manuals is obvious. It is perhaps also forgotten that documentation plays a major role in any design project where massive amounts of paper are used up reporting test results, explaining design rationales, documenting decisions and so forth. Electronic versions of this material might be extremely useful and cost effective. A design manager from British Aerospace (personal communication) estimated that 11–14 per cent of total costs in developing a new aeroplane went on producing and maintaining the documentation while a speaker at the 2nd UK Hypertext Conference (McAleese and Green, 1990) stated that code for a recent software program produced by British Telecommunications would run to 11 miles of paper if printed out in its entirety.

As yet, few electronic versions of either document type have been developed sufficiently to be used on a regular or widespread basis though it is likely that of the two, user manuals will emerge faster, if for no other reason than the economic interests underpinning technical manuals and their applications are more extensive than those of academic journal publishers.[2] Demonstration systems are emerging, particularly technical manuals for industrial applications, but little is currently known about their reception by users. Given the typical problems that are known to exist with electronic text however it is unlikely that rapid acceptance and use of either text type will occur.

6.4 Describing text usage

As mentioned in Chapter 3, process data of reading are hard to obtain reliably and in an unobtrusive manner. The standard approach of psychologists is to devise an experiment to answer any question. In the case of reading studies, this has led to a bias in favour of outcome measures such as speed or accuracy of reading performance, or comprehension level, although even the measurement of that particular outcome is the source of some debate. Unfortunately, where the classical experimental method (by which is meant the isolation, control, and manipulation of specified variables) proves difficult or impossible, it is often the problem that is considered ill-specified rather than any shortcomings of the formal experimental method that are exposed. This can be appreciated further by examining some of the more intrusive methods that reading experimentalists have employed to capture process data in a supposedly justifiable methodological manner. The use of bite bars and forehead rests (e.g. McConkie *et al.*, 1985) to hold readers' heads steady while text is passed before their eyes might sound sublime or ridiculous, depending on one's predilection, but there may be some reasonable doubts about their appropriateness for ecologically valid work.[3]

To insist on a laboratory-derived experimental approach to capturing process data here would not only require eye monitoring of the kind so beloved by legions of experimental psychologists but also demand a means of recording all physical manipulations of the text, possibly involving several video cameras, and an assurance from the readers that they would not move too fast or into a position that blocked out the camera's view. This is hardly a situation likely to induce relaxed examination of the reading process with interested readers and would therefore leave the current problem of how texts are read largely intractable thus lending support to Wittgenstein's (1953, p. 232) argument that in psychology: 'the existence of the experimental method makes us think that we have the means of solving the problems that trouble us; though problem and method pass one another by'.

The issues of why, what and how texts are read are our problems here, but they do not in themselves determine the data capturing methods to be employed in their solution. In the present context it was soon realized that the classic experimental method was not going to serve us well here although no single alternative technique offered the means to answer the questions being posed. However, the questions were considered legitimate therefore it was decided that a mixture of investigative procedures must be employed.

Interviewing a selection of relevant readers seems the most suitable means of gathering relevant data – why people read certain texts and what they typically expect the documents to contain. As mentioned previously, the advantages of interviewing are that it facilitates the elicitation of data that are difficult to obtain from more formal methods, as well as supporting the opportunistic pursuit of interesting issues. The problems and limitations of interviewing as a data elicitation method however, are well documented (e.g. Kerlinger, 1973). Common problems are the failure to structure the information gathering process properly so that certain topics are not asked of all interviewees or emphasis is placed on one topic at the expense of another. A further problem is the scoring or coding of what can be 'messy' data in a reliable manner. It is generally agreed however that potential shortcomings with interview techniques can be minimized by structuring the interview – following a fixed agenda – and using an agreed scoring scheme for data. The former is usually easier to derive than the latter.

In the work described here, the criteria derived from the repertory grid study outlined in the previous chapter provided a loose structure for eliciting and analysing the data. Thus a core set of issues to cover with every reader was identified and a standard means of categorizing answers was obtained. The interview questions were then devised by discussion, in the first instance, amongst the team of designers involved in building a journal database. This ensured that no important issues were overlooked and that the resultant data were of direct use to the design process. However, this also meant that for each investigation, the exact questions were never the same since they resulted from negotiations specific to the time and context of the investigation. This might be seen as source of weakness and in objective terms it may not be 'good science'. In defence of the approach though, the core questions (**why, what** and **how**) and their immediate derivatives were always addressed in a similar manner (as will be shown) and it is clear that the flexibility to add or alter some aspects of the investigation to suit the design situation could be seen more as a strength than a weakness of this approach, particularly in the unstable context of a software design process.

Interviewing alone though will not sufficiently answer the question of how a text is read. Certain issues or topics are not easy to describe adequately using only verbal means. To obtain suitable information in the present context it was decided that simulated usage or task performance with concurrent verbal protocols would complement the structured interview approach. The basic idea here was to ask readers to simulate their typical interaction with a text from the moment of first picking it up to the time of finishing with it, articulating what they were attending to with respect to the text at all times.

This method simulates probable task behaviour from the reader without resorting to elaborate means of investigation or generating masses of low-level data of the type certain to have emerged if readers had been set formal tasks and their interactions recorded on videotape. Readers in these investigations are asked to look at a selection of (predetermined) relevant texts and to examine each one as they normally would, e.g. if browsing them in the library, their office or whatever. They are prompted to articulate what information they cue into when they pick up a text, how they decide what is really worth reading and how they read say, journal articles that are selected for individual use. They repeat this simulation for several texts of that genre until a consistent pattern emerges at which point the reader is asked to confirm the accuracy of the investigator's interpretation of their reading style. In this way, the reader becomes part of the investigation procedure. What constitutes the description of the reading process is effectively negotiated between observer and reader, and the reader can at all stages correct the interpretations of the observer. The process concludes only when agreement is reached.

The main data source in this method is the verbal protocol. Like interviewing, much has been written about the use of verbal protocols in psychological investigations. The main issue of contention is the extent to which they can be said to reflect reliably the speaker's underlying cognitive processes or are merely a reflection of what the verbalizer thinks is appropriate and/or what they think the experimenter wants to hear (e.g. Nisbett and Wilson, 1977).

Ericsson and Simon (1984) have developed a framework for the use of verbal protocols and related it to current psychological theories of articulation which suggest that for tasks where subjects are required to describe what they are doing or attending to in real time (concurrent verbal reporting), objections on the grounds of inaccuracy or unreliability of self-reports rarely apply. Problems of accuracy are more likely to occur during retrospective verbal reporting ('This is how I did it ... '), as human memory is fallible and subject to post-task rationalization; or when reporting on how their own mental activities occurred ('I had an image of the object and then recalled ... '). In other words, when humans report what they are doing or trying to do during the performance (or simulation) of a task, and are not requested to interpret their own thinking (as in introspection), there are no *a priori* grounds for doubting the validity of their comments. Obviously subjects may lie or deliberately mislead the experimenter but this is a potential problem for all investigative methods requiring the subject to respond in a non-automatic fashion. The point here however is that people can reliably report what they are thinking – current contents of working memory – but are less likely to be able to do so for how they came to be thinking of it – what cognitive processes brought these contents to working memory.

Verbal protocol data of this form are regularly elicited in HCI studies and have been used to good effect in analysing the influence of various interface variables on users' perceptions of, and performance with a system (e.g. Mack *et al.*, 1983; Rasmussen, 1986; Dillon, 1987). Concurrent verbal protocols were used by the team at HUSAT not to provide deep insight into the cognitive process of the reader but to provide a verbal accompaniment to the behavioural act of text manipulation and usage. What is different from more typical verbal protocol analyses is the role of the verbalizer in clarifying their aims and behaviours.

To control for bias or potential limitations in the ability of one investigator to capture all relevant data another one was employed for the journal study (the first use of this approach) thus providing a co-rater for the elicited data and maximizing data capture. All interpretations and conclusions were checked with this experimenter also before final agreement was reached. After the first investigation, due to the high level of agreement between investigators and the lack of difficulty in one investigator recording the data with appropriate readers, this was felt to be unnecessary for further work. However, it would be wise for first-timers to work in pairs until familiarity with this form of data capture is gained.

6.5 Basic investigative method

Full details of the journal and software manual studies have been presented elsewhere – the journal analysis in Dillon *et al.* (1989); the software manual analysis in Dillon (1991a). In both cases the investigation was similar. Fifteen readers participated – this was originally deemed sufficient from the first study when it started to become clear that no new insights were being gained as more readers were being interviewed. It is possible that most major issues could be captured with less than this number but alternatively, it is also likely that a more accurate picture would emerge with many more participant readers.

Readers were interviewed individually and presented with a range of texts of the type under investigation. The basic procedure involved an interview to collect information on why readers used journals and what types of information they thought such a text contained. Typical prompts at this stage involved variations of the **why** and **what** aspects, pursuing points as they developed and concentrating on those areas that had any impact on usage.

Readers then interacted with a sample of relevant texts, simulating their typical usage according to their expressed reasons, articulating what they were attending to as they did so. They were prompted as necessary by the interviewer. After describing and simulating their typical usage of the texts the interviewer described his impression of their style and sought feedback from the reader that it concurred with the readers' own views. When agreement had been reached, i.e. the reader and interviewer both agreed that the representation of usage was accurate and adequate, the investigation ended.

6.6 Describing text usage according to Why, What and How attributes

To facilitate comparison between the two texts here and show how the general WWH approach operates, each attribute of usage is presented in turn and the data from both texts are combined.

6.6.1 Why read a text?

The most frequently stated reasons for accessing academic journals are summarized in Table 6.1 while the equivalent data for software manuals are presented in Table 6.2 and the data are rather self-explanatory. Virtually all readers distinguished between problem-driven journal usage, where work demands require literature reviews or rapid familiarization with a new area, and personal usage where journals are browsed in order to keep up with the latest developments in one's area of expertise or interest, the former being cited more frequently than the latter.

Table 6.1   Stated reasons for using journals

Why use a journal?	No. of Ss.	(%)
Background material for work purposes	11	(73)
Updating one's knowledge	7	(46)
Personal interest	3	(20)
On recommendation	2	(13)
Following up references	2	(13)

(Dillon *et al.*, 1989)

In more specific terms, journals are accessed for: personal interest, to answer a particular question (e.g. 'what statistics did the authors use and why?'), to keep up with developments in an area, to read an author's work and to gain advice on a research problem. In other words, there are numerous varied reasons for accessing material in journals apart from just wanting to 'study the literature'.

In terms of software manuals, readers stated numerous reasons for using manuals though there was a large degree of consistency in their responses.

Table 6.2   Stated reasons for using software manuals

Why use a manual?	No. of Ss.	(%)
For reference/How do I do this?	11	(73)
How to get started	10	(67)
When in trouble	8	(50)
For a summary of package's facilities	5	(33)
Aid for exploring software	2	(13)
As a guide before buying	1	(7)
For detailed technical info	1	(7)

(Dillon, 1991a)

The three main reasons (reference, introduction and when in trouble) were all offered by at least half the readers. These highlight the problem-driven nature of manual usage compared with the journals. In fact, all readers specifically remarked that manuals were only ever used in work or 'task' domains.

6.6.2 What type of information is in a text?

The text/graphics dichotomy

A major distinction seems to be drawn by readers between text and graphics in describing information types. The general consensus was that academic journals are a predominantly textual rather than graphical form of documentation (mentioned by 60 per cent) while all readers invoked this dichotomy in reporting that manuals relied too heavily on textual contents. Graphics in articles (tables, figures, etc.) were generally viewed positively as seven readers explicitly stated a dislike for academic articles that consisted of pages of straight text. A high proportion of mathematical content was also viewed negatively by the readers in this sample. More graphics would often aid the manual user's location and comprehension of information it was observed. However, it was repeatedly pointed out that graphics should be relevant and one manual (for the MacWrite package) was much criticized for using superfluous pictures of desktops, scissors and documents.

Linguistic style

The *linguistic* style was also mentioned in this context. The language of academic journals was seen to be relatively technical such that only readers versed in the subject matter could profitably read it. Furthermore, presentation style *was* seen as highly formalized, i.e. material written in a manner unique to journals that differs from conventional prose. In the case of software manuals, material was invariably described as 'technical', 'specific' and 'detailed'. While it might be argued that the very nature of manuals is that they contain such information, most readers seemed to find this off-putting. A third of the readers remarked that manuals were heavily loaded with jargon, with the result that information on simple actions was often difficult to locate or extract.

Information structure

The concept of structure in articles was discussed with all readers and it was apparent from their responses that most readers are aware of this as an organizing principle in documents. Organization of academic journal articles around a relatively standard framework was articulated, which the majority of the present respondents viewed as being of the form:

- Introduction
- Method
- Results
- Discussion/conclusion

or of the form:

- Introduction

- Elaboration and criticism of issues
- Alternative views proposed by author
- Discussion/conclusion

depending on whether it is an experimental or review type paper.

This order was seen as useful for reading purposes as it facilitated identification of relevant sections and allowed rapid decision making on the suitability of the article to a reader's needs (a hypothesis that I went on to formally test, see next chapter). For example, poor sectioning, large method and results sections, small discussions and large size in terms of number of pages were all cited as factors that would influence a reader's decision on whether or not to reject an article.

The structure of manuals was discussed with all readers and responses varied between those who are aware of text structure as it pertains to this text type and those who felt it existed but had difficulty articulating their perceptions of it. Primarily, a sense of order seems to be lacking in manuals though the majority (60 per cent) of readers felt that there might be a progression from easy to hard as a reader moves from the beginning to the end of the text, i.e. the more complex operations are dealt with towards the back of the manual. One reader remarked that a structure based around command frequency was probably common, i.e. frequently used commands or actions were more likely to be located at the front of the manual and less common ones at the back. Another suggested order, general-to-specific details, was made by two readers. Two readers argued that if any order such as easy-to-hard could be observed it probably existed at the task rather than the global level, i.e. within sections rather than across the manual.

The perceived modal structure for manuals that emerged in this study was

- contents
- getting started
- simple tasks
- more complex tasks
- index

As this structure indicates, heavy emphasis was placed on the task as a structural unit for organizing manuals. There were variations on this modal structure. For example, two readers placed training exercises at various points in the structure, the gradation between basic and more complex tasks was extended in two cases to include an intermediate level, while others mentioned a glossary of commands, technical specifications and lists of error messages as further typical units of a manual.

Many of the problems users of manuals seem to experience are related to the question of structure. Invariably this was criticized as 'poor' or 'disorganized'. The present sample seemed divided between those who felt that overall order was less important than the procedural order at the task level and those who were content with procedural ordering but felt that high-level ordering was unsatisfactory in many manuals. The need for different versions of manuals which are structured according to the users' needs was suggested by four readers. Typically it was suggested that these should consist of a manual for a 'total novice' which explains how to perform very basic procedures and a more detailed version for users who have acquired a greater degree of competence.

Physical size

The issue of document size is also interesting. Large academic articles obviously require a significant investment of time and this is often seen as a disincentive. Perceptions of what constituted a large or small article varied. Large articles were described as being anything from 6 to more than 30 pages long, medium length articles as being 5–20 pages long and small articles being 3–20 pages long. In other words what one individual rates as large, another may rate as small. Median responses suggest that articles more than 20 pages long are large and those articles that are about 5 pages long are small. Approximately 10 pages is considered to be medium length.

6.6.3 How are texts read?

Figure 6.1 represents a generic description of academic journal usage patterns. First, all readers skim read the table of contents of the journal issue. A preference was expressed for contents lists printed on the front or back page which made location of relevant articles possible without opening the journal. If the reader fails to identify anything of interest at this point the journal is put aside and, depending on the circumstances, further journals may be accessed and their contents viewed as above. When an article of interest is identified then the reader opens the journal at the start of the relevant paper. The abstract is usually attended to and a decision made about the suitability of the article for the reader's purposes.

At this point most readers reported also browsing the start of the introduction before flicking through the article to get a better impression of the contents. Here readers reported attending to the section headings, the diagrams and tables, noting both the level of mathematical content and the length of the article. Browsing the conclusions also seems to be a common method of extracting central ideas from the article and deciding on its relevance.

When readers have completed one initial cycle of interaction with the article they make a decision whether or not to proceed with it. A number of factors may lead to the reader rejecting the article. The main reason is obviously content and the accuracy of this first impression is an interesting empirical question.

If the article is accepted (or more likely, photocopied) for reading it is likely to be subjected to two types of reading strategy. The rapid scan reading of the article, usually in a non-serial fashion to extract relevant information, involves reading some sections fully and only skimming or even skipping other sections. The second reading strategy seemed to be a serial detailed read from start to

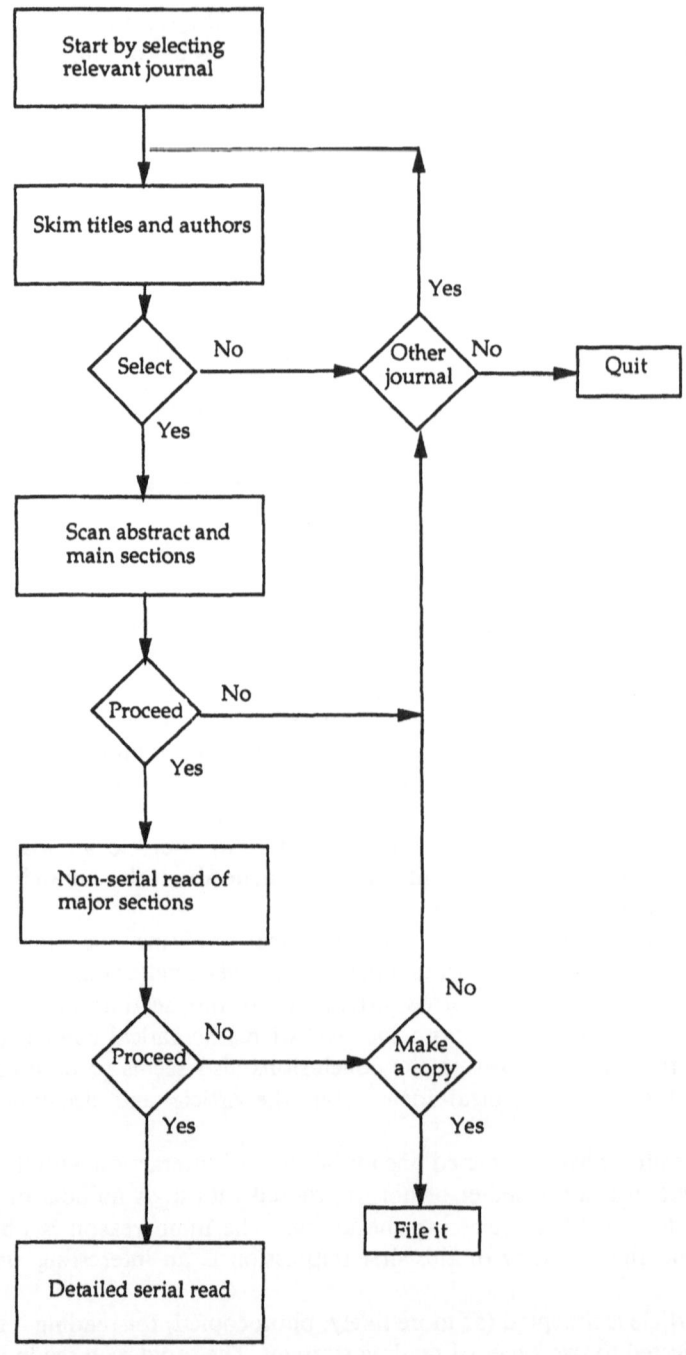

Figure 6.1 Generic model of journal use (after Dillon et al. *1989).*

finish. This was seen as 'studying' the article's contents and though not carried out for each article that is selected, most readers reported that they usually read selected articles at this level of detail eventually. Three readers expressed a preference for this reading strategy from the outset over scanning though acknowledging it to be less than optimum. While individual preferences for either strategy were reported most readers seem to use both strategies depending on the task or purpose for reading the article, time available and the content of the article.

Figure 6.2 represents usage styles for software manuals in a similar flowchart form. The first thing readers reported is 'getting a feel' for the

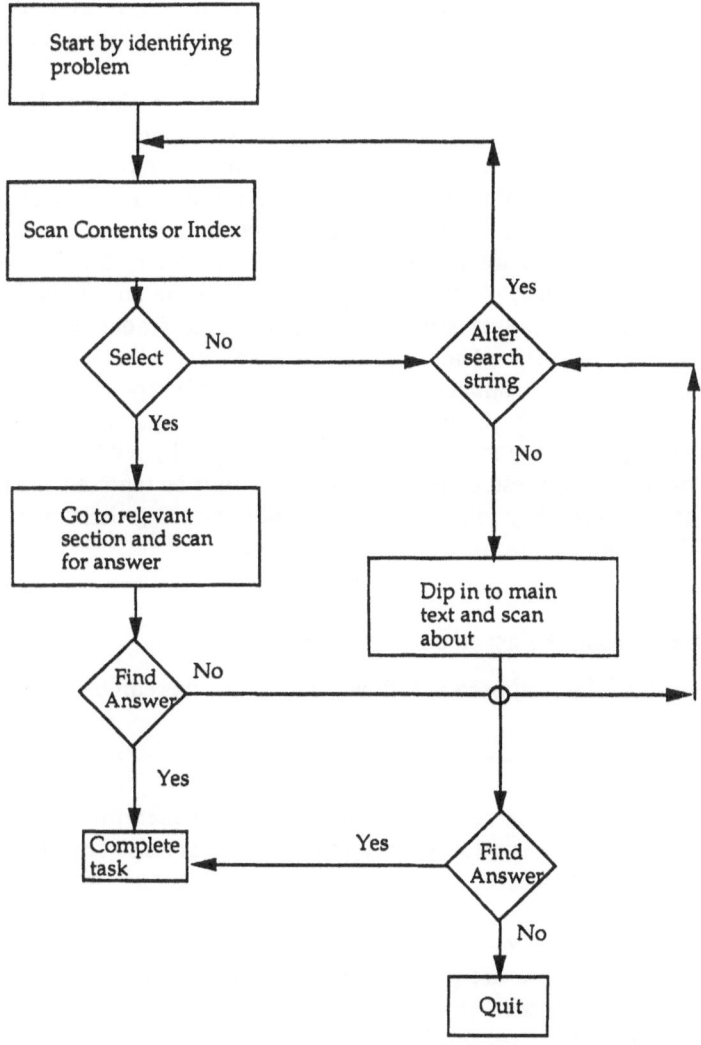

Figure 6.2   Generic model of manual usage (from Dillon, 1991a).

document's contents. Thus readers initially open the text at the contents or index sections. The majority (60 per cent) stated that the contents page was usually checked first and if that did not suggest where to go, the index was examined. However, it seems that much depends on the nature of the problem. If decoding an error message or seeking information on a particular command then the index is likely to provide this and will therefore be accessed first.

If more general information is sought then the contents offer a better chance of location. This highlights the extent to which book conventions have become internalized in the minds of contemporary readers. Furthermore, the index or contents list is not read in a simple pattern-matching fashion for a word template; rather the reader tries to get an impression of context from the contents, i.e. what items precede and proceed it, or may have to think of other terms that might yield satisfactory information if the term being sought is not present (a common problem with technical jargon).

If readers fail to locate anything in the contents or index that appears relevant they may either dip into the text and move about looking for relevant information or give up. The latter option appears common according to the data and represents a failure in document design that must be overcome.

If on the other hand a relevant item is located then the reader turns to the relevant page and if it is not immediately obvious, will scan about looking for a relevant diagram, word, phrase etc. to indicate that the answer is there.

At this stage the reading strategy adopted by the reader depends on the particulars of the task. The tendency to 'get on with it' seems firmly established in users of manuals and the present sample reported moving freely from manual to system in order to achieve their goal. Only three readers manifested any tendency to read around an area or fully read a section before moving on and even these admitted that they would be tempted to skim, and tend to get bored if they felt that they were not resolving their problems and only read complete sections if all else had failed.

6.6.4 Summarizing the usage data

In terms of the **WWH** approach followed here, the texts can be summarized as follows.

Academic journals

why: For work reasons such as keeping up with the literature, as a source of reference and as a source of learning. They are also read for personal reasons when accessed out of interest with no immediate work requirements.

what: Technical information about a specific domain; may have graphical components but are predominantly textual and tend to conform to a relatively standard structure.

how: Three levels of reading:

    1. quick scan of abstract and major headings;

   2.   non-serial scan of major sections;
   3.   full serial read of the text.

Software manuals

why: For task specific reasons such as troubleshooting, getting started, and for reference. Can occasionally be used for exploring software and identifying facilities or obtaining detailed technical information.

what: Technical information, of a specific and detailed nature, often laden with jargon. Can be a mixture of text and graphics. Structure is based around task units.

how: Problem driven. Broadly it involves checking the index or contents sections to find something relevant then dipping into and scanning sections of text. Lengthy serial reading is rare.

Thus, it is possible to gain information on the reading process for various texts using this approach and this information is more reliable than would be gained merely by self-analysis of one's own experiences with a document. In the following sections, the conclusions that were drawn from these studies for design are summarized.

6.7 Design implications

At the first level all readers attend to the contents page of journals and expect these to be easily accessible. It would seem therefore that a facility to scan lists of selectable titles and authors would be desirable. These should probably be grouped as they are on paper, i.e. in 'issues' although these could as easily be themes, but the ability to scan continually should be available.

Since the full contents of the paper are not attended to at the outset it would appear better that users are given brief information about the article and offered the chance of jumping around to various sections of the text. A likely presentation style based on these findings appeared to be: title, author(s), abstract, list of selectable section headings and the references cited. Further information about the size of the article might also be useful although how this is translated into meaningful metrics for electronic documents is unclear (does anyone really know how 'big' a 70 kbyte text file is?).

Rapid browsing facilities are vital. At this initial stage of article selection and the second level (non-serial read stage) fast page-turning is common as readers jump back and forth through the article. Any electronic version must support this activity by allowing both scrolling of the next page/previous page variety and rapid jumping to particular sections, e.g. from the introduction to the method or the discussion. It might be desirable to facilitate jumping to landmarks in the text such as tables or figures too, possibly with the use of a graphical browser.

The ability to print the article at any point would be desirable as obtaining hardcopies of selected articles appeared a major concern of these journal readers. Given the observed reading styles of the present sample, it appeared useful to consider offering the facility to print sections rather than the full article. For example, readers might choose to print the introduction and discussion sections only. This would have the advantage of reducing costs of obtaining hardcopies and save on unnecessary use of paper.

Potentially, the observed interrogative reading style for manuals seems highly suited to electronic presentations. However, an electronic text that merely replicates paper manuals would appear to be relatively useless. It would share all the disadvantages of the paper version but none of the advantages of manipulation, portability, familiarity and image quality. Since usage is so goal-oriented, large sections of a manual's contents are irrelevant for much of the time and providing a single multi-page text for the reader to search seems a less than optimum method of presenting information.

Therefore in order to ensure usability one would need to consider an alternative structure and means of access. Given the typical usage style noted above it is likely that searching facilities that supported location of information on a term or concept would be useful. It is also conceivable that a thesaurus of terms would enable better searching. Detailed contents/indexes with selectable items would be another enabling mechanism. None of these facilities should be beyond the scope of a well-designed hypertext application.

A continually emphasized attribute of hypertext is the ability of authors to hide layers of information that can be accessed by links. Thus users could conceivably follow a trail of information about a concept to whatever level of detail is required. For the curious user this could be easily achieved by actioning links. For the user who has no desire for such in-depth coverage such a presentation format could hide copious amounts of unwanted material. Thus we can see the emergence through templates of versions of manuals for different user types as suggested by some readers. Such a structure would support certain aspects of the typical model which readers possess of this information type.

Obviously this discussion of potential hypertext attributes to support electronic manuals is superficial. No mention has been made of the potential for such attributes as 'interactive figures' (where users can point at sections of a diagram to gain further information on that part), scrollable pop-up windows or links to alternative procedures and methods for accomplishing a goal. Intelligent searching facilities that provided links to associated concepts when a specific term failed to elicit a satisfactory response would obviously be useful but is beyond the scope of consideration for the present book.

What this study has shown however, is that the classification criteria can also be applied to another text type to support the elicitation of specific information on usage. Given the distinctions made between journals and manuals in the repertory grid analysis of Chapter 4, there is reason to believe that this form of analysis could usefully be applied to a wide range of text types.

6.8 General conclusions and the way forward

After reviewing the literature on reading from screens it was stated that providing designers of electronic text with guidance on the basis of experimental evidence was not straightforward. The lack of a suitable description of the reading process was highlighted as a major problem in this venture. It was recognized that some form of text classification would be a useful starting point for such a description and the analysis of reader's classifications in Chapter 5 provided three criteria for distinguishing texts. In the present chapter a demonstration of the application of these criteria to two texts has been provided.

Though probably a non-optimum means of gathering reading process data, this simulation-based procedure demonstrates that it is at least possible to gain design-relevant information on the reading process without recourse to sophisticated and/or unnecessarily intrusive tools. It has been my experience that readers have few difficulties articulating **how** they use texts and responding to questions on the **what** and **why** aspects of reading. There is normally a high degree of consistency between readers which enables general conclusions to be drawn, yet the data clearly distinguish between text types in the characterization of usage they suggest.

Data of this type are useful for gaining insight into the relevant design issues for an electronic document, particularly early in the product life cycle (a time when human factors inputs are typically weak) but there are limitations. It is applicable in the main, only for existing texts. For innovative information types that will surely emerge with the advent of hypermedia such a classification of task relevant issues is not possible (although subsets of the hypermedia information might be amenable to such analysis). Furthermore it is a relatively informal procedure, reliant for its value ultimately on the abilities of the practitioner more than one might like, particularly for mapping responses to interface recommendations.

However, this form of data capture and analysis is not intended to be anything more than an accurate means of initially conceptualizing the issues, i.e. identifying the text type and its associated usage characteristics. By providing designers with some reliable estimates of the three aspects of usage – the **why**, **what** and **how** of reading – it supports reasoned constraining of the design options under consideration. More formal analyses can then be appropriately targeted at specific design issues for the artefact under consideration.

As an investigative procedure, it is quick to perform and seems unlikely to tax readers unduly. For sure it does not reveal where the reader's eyes go or accurately map the sequence of pages followed but in defence, one must ask if that is that what is needed at this stage? Approximate data of these kind offer a means of constraining the space of design options and highlighting aspects of reading that the technology could improve. In so doing, it involves intended users of the technology and offers a means of negotiating potential solutions with them.

Notes

1. Copyright is a major concern in this area as publishers seek ways to control delivery and copying of material in electronic form. It will not be discussed in detail in the present work though it will be referred to where it places practical constraints on any recommendations made on the basis of ergonomics.

2. That said, electronic journals ('ejournals' to their friends!) are springing up regularly, though few have made it as far as competing with their paper equivalents in terms of professional status. Interestingly enough, it is not the ergonomic issues that are hotly debated in discussing the pros and cons of ejournals but the problem of professional recognition and citation for articles published in this medium.

3. The term 'ecologically valid' describes work that seeks to reflect the real-world aspects of a task rather than work that takes the issue under investigation out of its natural situation of occurrence and places it in isolation. It is a term that causes a lot of debate amongst human scientists not least in the present context. A recent article of some colleagues and mine was criticized for using the expression to describe a measure we had taken. The referee never explained their objection but merely wrote 'Ecological validity? Not in this journal please'!

7

Information as a structured space

The relationship between a writer and a reader is a sacred space. No one
has the capacity to mess with my words.

Marianne Williamson, interview in *Empire* (Dec. 1992)

7.1 Introduction

As stated earlier, the emergence of electronic document forms such as
hypertext, make it possible to embody alternative and multiple structures for
electronic texts that could not be supported feasibly in the comparatively
standard format of paper. Typically, advocates of the 'new structures'
approach dismiss paper as a limiting medium, claiming it imposes a 'linear
strait-jacket of ink on paper' (Collier, 1987). This is contrasted with the
supposedly liberating characteristics of hypertext, which, with its basic form
of nodes and links, is seen to be somehow freer or more natural. Nielsen (1990,
p. 1) for example, would appear to reflect the consensus view with the
following summary:

> All traditional text, whether in printed form or computer files, is
> sequential, meaning that there is a single linear sequence defining the
> order in which it is to be read. First you read page one. Then you read
> page two. Then you read page three. And you do not have to be much of a
> mathematician to generalize the formula which determines which page to
> read next. . . . Hypertext is nonsequential; there is no single order in
> which the text is to be read.

Similar and even more extreme positions can be found in many of the
(interestingly, paper-based) writings describing the wonders of hypertext
(Beeman *et al.* 1987; Nelson, 1987; Landow, 1991, *inter alia*).

One may not need to be a sophisticated mathematician to derive a formula
to predict such a reading style but one would be a poor student of human
psychology if one really believed any resulting formula provided an accurate
representation of the reading process. Certainly reading a sentence is largely a
linear activity (although eye-movement data suggest that even here, non-

linearity can occur) so even on screen, reading of sentences is still as likely to be predominantly linear. The issue then is the extent to which linearity is imposed on the reader at the above-sentence level in using the document.

Though paper texts may be said to have a physical linear format there is little evidence to suggest that readers are constrained by this or only read such texts in a straightforward start-to-finish manner (Pugh, 1975; Charney, 1987; Horney, 1993). For example, the journal usage study (see Chapter 6) identified three reading strategies in readers of academic journals, only one of which could be described as linear, and one only has to think of one's own typical interaction with a newspaper to demolish arguments of constrained linear access and talk of 'strait-jackets'. With hypertext, movement might involve less physical manipulation (a keypress for example compared to turning and folding pages) or less cognitive effort (a direct link to information compared to looking up an index and page number) but these are matters of degree, not of category. One could make a case for paper being the liberator as at least the reader always has access to the full text (even if searching it might prove awkward). With hypertext, the absence of links could deny some readers access to information and always forces them to follow someone else's ideas of where the information trail should lead.

Although the notion of linearity is much abused in this domain, the notions of access paths and structure have given rise to the idea of information as having shape and occupying space through which a reader must navigate. In seeking to provide innovative and/or natural structures for information, the question of navigation has become a central one to the field. In the present chapter, these ideas are examined and the extent to which text designers must address such issues is considered.

7.2 The concept of structure in documents

Regardless of the putative constraints of paper texts or browser-friendly attributes of hypertexts, it seems certain that readers possess some knowledge of a document type that provides information on the probable structure and organization of key elements within it. For example, the moment we pick up a book we are afforded numerous cues that as experienced readers, we are likely to respond to such as size (which can be an indicator of likely effort involved in reading it), age or condition (which can act as both an indicator of relevance as well as of previous usage rates). Such affordances give at least partial lie to the cliché that 'you cannot judge a book by its cover'; you can, and you do.

Once the book is opened, still more cues are available to readers, informing them of where material is located, how it is organized and what information is included. This information is available through contents lists, chapter headings, abstracts, and summaries. The regular reader comes to expect this information and probably picks it up in a relatively automatic fashion when browsing a book. Almost certainly, we would think it odd if the contents list was missing or the index was at the front. We expect certain consistency in layout and structure and notice them most when this expectation is violated.

The same point might be argued for a newspaper. Typically we might expect a section on the previous day's political news at home, foreign coverage, market developments and so forth. News of sport will be grouped together in a distinct section and there will also be a section covering that evening's television and radio schedules. With many paper documents there tend to be at least some standards in terms of organization so that concepts of relative position in the text such as 'before' and 'after' have tangible physical correlates. If this can be said to hold true for all established text forms, then developers of hypertext systems need to consider carefully their designs in terms of whether they support or violate such assumptions.

Unfortunately, the term 'structure' is used in at least three distinct ways by different researchers and writers in this field. Conklin (1987) talks of structure being imposed on what is browsed by the reader, i.e. the reader builds a structure to gain knowledge from the document. Trigg and Suchman (1989) refer to structure as a representation of convention, i.e. it occurs in a text form according to the expected rules a writer follows during document production. Hammond and Allinson (1989) offer a third perspective, that of the structure as a conveyor of context. For them, there is a naturally occurring structure to any subject matter that holds together the 'raw data' of that domain and supports reading.

In reality, there is a common theme to all these uses. They are not distinct concepts sharing the same name but different affects or manifestations of the same concept. The main role of structure seems to differ according to the perspective from which it is being discussed: the writer's or the reader's, and the particular part of the reading/writing task being considered. Thus the structure of a document can be a convention for both the writer, to conform to expectations of format, and for the readers, so they know what to expect. It can be a conveyer of context mainly for the reader to infer from, and elaborate on, the information provided, but it might be employed by a skilled writer with the intention of provoking a particular response in the reader. Finally, it can be a means of mentally representing the contents both for the readers to grasp the organization of the text and for the author appropriately to order this delivery.

From the comments of readers in the journal and manual usage studies (see Chapter 6) it is clear that structure is a concept for which the meanings described above seem to apply with varying degrees of relevance. Certainly the notion of structure as convention seems to be perceived by readers of journal articles while the idea of structure supporting contextual inference seems pertinent to users of software manuals. Recent evidence from interviews with readers suggests they are aware of the presence of structure in several other text forms such as manuals, magazines and novels. Nijaar (1993) interviewed 20 readers about their perceptions of structure in documents and observed that the majority were aware of conventional forms for such texts as magazines, newspapers and academic journals but were less able to articulate exactly what form these conventions took. Beyond these manifestations, research in the domain of linguistics and discourse comprehension lends strong support to the concept of structure as a basic component in the reader's mental representation of a text.

The van Dijk and Kintsch (1983) theory of discourse comprehension places great emphasis on text structure in skilled reading. According to their theory, reading involves the analysis of propositions in a text and the subsequent formation of a macropropositional hierarchy (i.e. an organized set of global or thematic units about the events, acts, and actors in the text). From this perspective, increased experience with texts leads to the acquisition of knowledge about macrostructural regularities which van Dijk and Kintsch term 'superstructures' that facilitate comprehension of material by allowing readers to predict the likely ordering and grouping of constituent elements of a body of text in advance of reading it. They postulate three general levels of text unit: microstructures, macrostructures and superstructures but prefer to concentrate on the first two, at this time having developed their ideas on these to a greater extent than they have on the third. However, experimental work seems to confirm the relevance of the third level of structure even if its exact relationship to their comprehension theory is not precisely specified yet.

They have applied this theory to several text types. For example, with respect to newspaper articles they describe a schema consisting of headlines and leads (which together provide a summary), major event categories each of which is placed within a context (actual or historical), and consequences. Depending on the type of newspaper – weekly/daily, tabloid/quality – one might expect elaborated commentaries and evaluations. Experiments by Kintsch and Yarborough (1982) showed that articles written in a way that adhered to this schema resulted in better grasp of the main ideas and subject matter (as assessed by written question answering) than ones which were reorganized to make them less schema conforming.

Interestingly, when given a cloze test of the articles, i.e. a traditional comprehension test for readers that requires them to fill in the blanks within sentences taken from the text they have just read, no significant difference was observed. The authors explain this finding by suggesting that schematic structures are not particularly relevant as far as the ability to remember specific details such as words is concerned (i.e. the ability which is measured by a cloze test) but have major importance at the level of comprehension. In their terms, word processing and recall is handled at the microstructural level, text specific organization at the macrostructural level and general organization of the text type at the superstructural level.

The van Dijk and Kintsch theory has been the subject of criticism from some cognitive scientists. Johnson-Laird (1983) for example takes exception to the idea of any propositional analysis providing the reader with both the basic meaning of the words in the text and the significance of its full contents. For him, at least two types of representational format are required to do this and he provides evidence from studies of people's recall of text passages that it is not enough to read a text correctly (i.e. perform an accurate propositional analysis) to appreciate the significance of that material. He proposes what he terms mental models as a further level of representation that facilitates such understanding. Subsequent work by Garnham (1987) lends support to the insufficiency-of-propositions argument in comprehension of text (though neither have directly tackled van Dijk and Kintsch's third level, the super-structural).

The differences between Johnson-Laird and van Dijk are mainly a reflection of the theoretical differences between the psychologist's and the linguist's views of how people comprehend discourse. From the perspective of the human factors practitioner it is not clear that either theory of representation format is likely to lead to distinct (i.e. unique) predictions about electronic text. Both propose that some form of structural representation occurs – it is just the underlying cognitive form of this representation that is debated. The similarity of their views from the human factors perspective is conveyed in this quote from Johnson-Laird (1983, p. 381) where he states that mental models:

> appear to be equally plausible candidates for representing the large-scale structure of discourse – the skeletal framework of events that corresponds to the 'plot of the narrative', the 'argument' of a non-fiction work and so on. Kintsch and van Dijk's proposal that there are macrorules for constructing high-level representations could apply *mutatis mutandis* to mental models.

In other words, the issue is not if, or even how, readers acquire a structural representation of texts they read (these are accepted as givens) but what form such structures take: propositions or mental models? van Dijk and Kintsch (1983) addressed some of the Johnson-Laird criticisms by incorporating a 'situation model' of the text into their theory. This is a mental representation of the significance of the text in terms of its subject matter and the central figures/elements under discussion which facilitates the application of contextual knowledge stored in long-term memory (a weakness of their original proposition based theory).

The debate on representational format largely side-steps the structural issue of most relevance to electronic text design, the high-level organization of presented material. This is van Dijk and Kintsch's third level, the superstructural, which is akin to the text conventions and organization principles not directly tied to the reader's attempts at comprehension through cognitive representation of the message. As yet, we have little theoretical debate at this level. For the human factors practitioner, theoretical distinctions of supposed representational form are unlikely to be as important to the design of electronic documents as consideration of the physical text structure that gives rise to the cognitive experience and the concomitant so-called 'representations'. The major issue in this context therefore is the extent to which structure is perceived in certain texts and how designers must accommodate this in their products.

7.3 Schema theory as an explanatory framework

In viewing structure as a component of texts the conceptualization emerges of information as space and the reader as a navigator. This in turn invites a mapping between psychological theory and information design that is frequently accepted by researchers in this domain. Schema theory provides a convenient explanatory framework for the general knowledge humans seem to possess of activities, objects, events and environments. It postulates the existence of some form of general knowledge of the world that aids humans in handling information impinging on their senses and offers information designers a parsimonious theoretical account of the information structure and reader navigation issues. For a fuller discussion of this account, see Dillon *et al.* (1993b).

It seems obvious, for example, that we must possess schemata of the physical environment we find ourselves in if we are not to be overwhelmed by every new place we encounter. Presumably acquired from exposure to the world around us, schemata can be conceptualized as affording a basic orienting frame of reference to the individual. Thus, we soon acquire schemata of towns and cities so that we know what to expect when we find ourselves in one – busy roads, numerous buildings, shopping areas, people, etc.

In employing schema theory as our explanatory framework it is worth making a distinction between what Brewer (1987) terms 'global' and 'instantiated' schemata. The global schema is the basic or raw knowledge structure. Highly general, it does not reflect the specific details of any object or event (or whatever knowledge type is involved). The instantiated schema however is the product of adding specific details to a global schema and thereby reducing its generality. An example will make this clearer. In orienting ourselves in a new environment we call on one or more global schemata (e.g. the schema for city or office building). As we proceed to relate specific details of our new environment to this schema we can be said to develop an instantiated schema which is no longer general but is not sufficiently complete to be a model or map of the particular environment in which we find ourselves.

Global schemata remain to be used again as necessary. Instantiated schemata presumably develop in detail until they cease to be accurately described as schematic or are discarded when they serve no further purpose. In the above example, if we leave this environment after a short visit we are likely to discard the instantiated schema we formed, but if we stay or regularly return we are likely to build on this until we have detailed knowledge of the environment.

While schemata are effective orienting guides, in themselves they are limited. In particular, they fail to reflect specific instances of any one environment and provide no knowledge of what exists outside our field of vision. As such, they provide the basic knowledge needed to interact with an environment but must be supplanted by other representations if we are to plan routes, avoid becoming lost or identify short cuts – activities in which humans seem frequently to engage.

7.3.1 Levels of schema instantiation: landmarks, routes and surveys

The first postulated stage of schema instantiation is knowledge of 'landmarks', a term used to describe any features of the environment which are relatively stable and conspicuous. Thus we recognize our position in terms relative to these landmarks, e.g. our destination is near a large, new building or if we see a statue of a soldier on horseback then we must be near the railway station and so forth. This knowledge provides us with the skeletal framework on which we build our cognitive map.

The second stage of instantiation is route knowledge which is characterized by the ability to navigate from point A to point B, using whatever landmark knowledge we have acquired to make decisions about when to turn left or right. With such knowledge we can provide others with effective route guidance, e.g. 'Turn left at the traffic lights and continue on that road until you see the large church on your left and take the next right there . . .' and so forth. Though possessing route knowledge, a person may still not really know much about their environment since any given route might be effective but non-optimal or even totally wasteful.

The third stage of instantiation is survey (or map) knowledge. This allows us to give directions or plan journeys along routes we have not directly travelled as well as describe relative locations of landmarks within an environment. It allows us to know the general direction of places, e.g. 'westward' or 'over there' rather than 'left of the main road' or 'to the right of the church'. In other words it is based on a world frame of reference rather than an egocentric one.

Current thinking is dominated by the view that landmark, route and survey knowledge are points on a continuum rather than discrete forms. The assumption is that each successive stage represents a developmental advance towards an increasingly accurate or sophisticated world view. Certainly this is an intuitively appealing account of our own experiences when coming to terms with a new environment or comparing our knowledge of one place with another and has obvious parallels with the psychological literature which often assumes invariant stages in cognitive development, but it might not be so straightforward.

Obviously landmark knowledge on its own is of little use for complex navigation and both route and survey knowledge emerge from it as a means of coping with the complexity of the environment. However it does not necessarily follow that given two landmarks the next stage of knowledge development is acquiring the route between them or that once enough route knowledge is acquired it is replaced by or can be formed into survey knowledge. Experimental investigations have demonstrated that each form of representation is optimally suited for different kinds of tasks (e.g. Thorndyke and Hayes-Roth, 1982; Wetherell, 1979). Route knowledge is cognitively simpler than survey knowledge but suffers the drawback of being virtually useless once a wrong step is taken (Wickens, 1984). Thus while the knowledge forms outlined here are best seen as points on a continuum and a general trend to move from landmark to survey knowledge via route knowledge may exist, task dependencies and cognitive ability factors mediate such developments and suggest that an invariant stage model may not be the best conceptualization of the findings.

7.3.2 Documents as navigable structures

If picking up a new book can be compared to a stranger entering a new town (i.e. we know what each is like on the basis of previous experience and have expectancies of what we will find even if we do not yet know the precise location of any element) how do we proceed to develop our map of the information space?

To use the analogy of the navigation in physical space we would expect that generic structures such as indices, contents, chapter headings and summaries can be seen as landmarks that provide readers with information on where they are in the text, just as signposts, buildings and street names aid navigation in physical environments. Thus when initially reading a text we might notice that there are numerous figures and diagrams in certain sections, none in others, or that a very important point or detail is raised in a section containing a table of numerical values. In fact, readers often claim to experience such a sense of knowing where an item of information occurred in the body of the text even if they cannot recall that item precisely and there is some empirical evidence to suggest that this is in fact the case.

Rothkopf (1971) tested whether such occurrences had a basis in reality rather than resulting from popular myth supported by chance success. He asked people to read a 12-page extract from a book with the intention of answering questions on content afterwards. What subjects did not realize was that they would be asked to recall the location of information in the text in terms of its occurrence both within the page and the complete text. The results showed that incidental memory for locations within any page and within the text as a whole were more accurate than chance, i.e. people could remember location information even though they were not asked to. There was also a positive correlation between location of information at the within-page level and accuracy of question answering. Work by Zechmeister *et al.* (1975), McKillip *et al.* (1975) and Lovelace and Southall (1983) confirm the view that memory for spatial location within a body of text is reliable even if it is generally limited. Simpson (1990) has replicated this for electronic documents.

In the paper domain at least, the analogy with navigation in a physical environment is of limited applicability beyond the level of landmark knowledge. Given the fact that the information space is instantly accessible to the reader (i.e. they can open a text at any point) the necessity for route knowledge, for example, is lessened (if not eliminated). To get from point A to point B in a text is not dependent on taking the correct course in the same way that it is in a physical three-dimensional environment. The reader can jump ahead (or back), guess, use the index or contents or just page serially through. Readers rarely rely on just one route or get confused if they have to start from a different point in the text to go to the desired location, as would be the case if route knowledge was a formal stage in their development of navigational knowledge for texts. Once you know the page number of an item you can get there as you like. Making an error is not as costly as it is in the physical world either in terms of time or effort. Furthermore, few texts are used in such a way as to require that level or type of knowledge.

One notable exception to this might be the knowledge involved in navigating texts such as software manuals or encyclopædias which can consist of highly structured information chunks that are inter-referenced. If, for example, a procedure for performing a task references another part of the text, it is conceivable that a reader may only be able to locate the referenced material by finding the section that references it first (perhaps because the index is poor or they cannot remember what it is called). In this instance one could interpret the navigation knowledge as being a form of route knowledge. However, such knowledge is presumably rare except where it is specifically designed into a document as a means of aiding navigation along a trouble-shooting path.

A similar case can be made with respect to survey knowledge. While it seems likely that readers experienced with a certain text can mentally envisage where information is in the body of the text, what cross-references are relevant to their purpose and so forth, we must be careful that we are still talking of navigation and not changing the level of discourse to how the argument is developed in the text or the ordering in which points are made. Without doubt, such knowledge exists, but often it is not purely navigational knowledge but an instantiation of several schemata such as domain knowledge of the subject matter, interpretation of the author's argument, and a sense of how this knowledge is organized that come into play now. This is not to say that readers cannot possess survey type knowledge of a text's contents, rather it is to highlight the limitations of directly mapping concepts from one domain to another on the basis of terminology alone. Just because we use the term navigation in both situations does not mean that they are identical activities with similar patterns of development. The simple differences in applying findings from a three-dimensional world (with visual, olfactory, auditory and powerful tactile stimuli) to a two-dimensional text (with visual and limited tactile stimuli only) and the varying purposes to which such knowledge is put in either domain are bound to have a limiting effect.

It might be that rather than route and survey knowledge, a reader develops a more elaborate analogue model of the text based on the skeletal framework of landmark knowledge outlined earlier. Thus, as familiarity with the text grows, the reader becomes more familiar with the various landmarks in the text and their inter-relationships. In effect, the reader builds a representation of the text similar to the survey knowledge of physical environments without any intermediary route knowledge but in a form that is directly representative of the text rather than physical domain. This is an interesting empirical question and one that is far from being answered by current knowledge of the process of reading.

7.3.3 Schemata and electronic space

The concept of a schema for an electronic information space is less clear-cut than for physical environments or paper documents. Computing technology's short history is one of the reasons for this but it is also the case that the medium's underlying structures do not have equivalent transparency. With paper, once the basic *modus operandi* of reading is acquired (e.g. page-turning, footnote identification, index usage and so forth) it retains utility for other texts produced by other publishers, other authors and for other domains. With computers, manipulation of information can differ from application to application within the same computer, from computer to computer and from this year's to last year's model. Thus using electronic information is often likely to involve the employment of schemata for systems in general (i.e. how to operate them) in a way that is not essential for paper-based information.

The qualitative differences between the schemata for paper and electronic documents can easily be appreciated by considering what you can tell about either at first glance. I have outlined the information available to paper text users in the section on paper schemata above. When we open a hypertext document, however, we do not have the same amount of information available to us. We are likely to be faced with a welcoming screen which might give us a rough idea of the contents (i.e. subject matter) and information about the authors/developers of the document but little else. It is two-dimensional, gives no indication of size, quality of contents, age (unless explicitly stated) or how frequently it has been used (i.e. there is no dust or signs of wear and tear on it such as grubby fingermarks or underlines and scribbled comments). At the electronic document level, there is usually no way of telling even the relative size without performing some 'query operation' which will usually return only a size in kilobytes which conveys little meaning to the average reader.

As outlined in Chapter 3, performing the hypertext equivalent of opening up a document or turning the page offers no assurance that expectations will be met since many hypertext documents offer unique structures. At their current stage of development it is likely that users/readers familiar with hypertext will have a very limited or poorly developed schema. The manipulation facilities and access mechanisms available in hypertext will probably occupy a more prominent rôle in their schema for hypertext documents than they will for readers' schemata of paper texts since in the electronic domain they currently differ from application to application.

The fact that hypertext offers authors the chance to create numerous structures out of the same information is a further source of difficulty for users or readers. Since schemata are generic abstractions representing typicality in entities or events, the increased variance of hypertext implies that any similarities that are perceived must be at a higher level or must be more numerous than the schemata that exist for paper texts. It seems therefore that users' schemata of hypertext environments are likely to be 'informationally leaner' than those for paper documents. This is attributable to the recent emergence of electronic documents and comparative lack of experience interacting with them as opposed to paper texts for even the most dedicated users. The current lack of standards in the electronic domain compared to the rather traditional structures of many paper documents is a further problem for schema development.

7.4 Identifying structural forms in electronic document design

As mentioned previously, the author was a participant in a team designing a hypertext database of journal articles. It was obvious from the literature, from theoretical considerations and from the evidence of the BLEND and ADONIS systems that straight reproduction of the paper journal format was unlikely to prove acceptable in usability terms, even if more advanced manipulation facilities had emerged since either of those systems were specified.[1]

The question of how such texts might be structured to maximize usability was of direct concern to this design project. Furthermore, the journal usage study cited earlier had shown that typical usage of this text type involved jumps and non-serial access routines for which a model of an article's typical structure was a useful guide. Thus there were two apparently conflicting requirements: support non-serial access while retaining a supposedly useful linear structure.

The investigations reported here marked an attempt to clarify these issues in the context of the database under design. The major issue was to identify the relevance of the article superstructure to the database in question. Given that this design scenario offered the potential to restructure journal articles radically, how should we identify relevant or possibly better structures? Since the first law of ergonomics is 'know thy user' it was decided to run a quick series of experiments to let readers propose suitable structures.

7.5 Examining structure experimentally

The specific aims of this experimental work were to identify the extent to which readers possessed any superstructural representation or model of a typical academic article and to examine how it might be transferred to, or affected by screen presentation. A synopsis of these experiments is provided here; full details may be found in Dillon (1991b).

7.5.1 Readers, texts and tasks

In both experiments readers who participated were typical of the intended users of the hypertext database. All were professional researchers experienced in the use of academic articles and were habitual users of Apple Macintosh computers.

Articles were selected from one journal in the researchers' field of expertise. The articles were matched approximately for size and number of paragraphs, presence of figures and tables, and conformation to the single experiment report style. Though roughly in the area of interest to the researchers they were selected so as not to have been read by these participants.

In both experiments, readers were presented with parts of the text and asked to organize them in terms of overall structure as quickly as they could without concerning themselves with comprehension of the material. They knew only that the material was from an academic article in their subject domain. Instructing them to work quickly and not to concern themselves with comprehending the material was intended to prevent organization by problem-solving or lengthy reasoning and to accentuate initial impressions of suitable structures.

In the first experiment, the texts consisted of a selection of paragraphs, figures and tables from two academic journal articles. In a repeated measures design, readers attempted to assemble two articles, one with relevant headings present and the other without headings. To avoid referential continuity, every second paragraph was removed from the original text.

In the second study, readers read a selection of paragraphs from two similarly selected articles but this time were asked only to indicate each paragraph's relative location in terms of general sections identified in the first experiment. Readers performed the task both on paper and with electronic versions of the two texts.

7.5.2 Results from experiment 1

Despite the low levels of absolute accuracy (mean rate only 16.7 per cent, no effect for headings) it was clear that all readers were imposing (as required) a structure on each article of the form: introduction/method/results/discussion (hereafter referred to as the IMRD format). Analysing their assemblies in these terms, it emerged that much higher general accuracy levels were present (mean accuracy rate = 82.58 per cent) indicating that readers can predict location of isolated paragraphs of text in their correct general sections with high levels of accuracy. Once more, no effect was observed for headings. Speed of performance indicated that times in the No Headings condition (mean time = 718 s., s.d. = 162 s.) were slightly (but not statistically significantly) faster than those in the Headings condition (mean time = 783 s., s.d. = 175 s.).

Apart from the general accuracy levels observed, it was interesting to note the type of mistakes made by readers performing this task. Three basic errors were identified:

1. secondary heading placement;
2. figure and table placement; and
3. introduction/discussion distinction.

The most obvious problem occurred with the secondary headings. Primary headings (Introduction, Results, etc.) were easily placed but these are relatively standard, secondary headings tend to be unique to the article, reflecting the author's views of a section's contents. For example, a heading such as 'The Effect of Display Size' might fit logically into the results section when read in context but taken as an isolated piece of text could as easily be a heading in the introduction or discussion sections of an academic article.

Figures and tables posed problems in terms of absolute accuracy too, although subjects usually placed these in the correct section. This is not too difficult to explain, their occurrence in articles of this form is rare outside of the results section. Non-graph/numerical types might pose more of a problem but even they are unlikely to occur in introduction/discussion sections.

A common error was the confusion of introduction and discussion paragraphs. All subjects made this mistake at least once. In terms of the type of text usually found in these sections this is understandable. Both contain general text with references to other related work, a form atypical of other sections. Thus while it is easy to identify isolated paragraphs as belonging to these sections, it is less easy to distinguish correctly between them.

All readers were required to describe briefly (in written form) the contents of the text they had just assembled. Of the dozen readers participating, ten remarked that they had little memory of the text and had not read it for comprehension (as instructed). While it is interesting that they could assemble the text without reading it for comprehension purposes, all readers were capable of providing a rough sketch of the article. Typically they accurately reported the subject matter, that it was an experimental paper, the type of design or analysis, and its broad aims. In some cases parts of the results or their implications were also grasped. There were numerous inaccuracies however and most of the written reports were in the form of keywords or short phrases suggesting little attempt to grasp the development of the argument within the text.

It was clear from this study that these readers possessed some knowledge (presumably schematic) of the text's structure independent of its semantic content. However, much work in the area of electronic text has shown that when text is presented on screen many of the findings from the paper domain cease to hold. The present study therefore sets out to examine the ability of readers to predict location by applying the superstructural representation of an article to information on screen.

7.5.3 Results from experiment 2

In the screen condition for this experiment paragraphs were presented mid-screen as black text on a totally white background using HyperCard on an Apple Macintosh Plus. The only other information present was the number of the paragraph (1–20) in the top right corner and a 'button' in the lower centre of screen facilitating movement to the next card. In the paper condition paragraphs were presented on 20 sheets of paper printed from this HyperCard stack, of similar size to the screen and stapled together in the top left corner. They contained identical information except for the 'button'.

Mean performance time with paper was significantly faster than with screen presented text (paper mean/s.d. = 234/99 s.; electronic mean/s.d. = 280/84 s.; $t = 3.16$, df = 7, $p < 0.02$). The mean number of errors per reader was similar for each presentation medium although three-quarters of the readers performed better or as well with the electronic text (paper mean/s.d. = 3.88/0.64 errors; electronic mean/s.d. = 3.5/3.07 errors).

Overall accuracy levels were similar to experiment 1, 81.55 per cent for combined conditions, 80.6 per cent for paper alone, 82.5 per cent for screen alone, confirming the earlier finding that the ability to predict location on the basis of limited information is highly developed for experienced readers of this text type.

Twelve types of error could be made (number of categories × number of incorrect categories per item). In total, 59 errors were made. These are summarized in Table 7.1.

Table 7.1   Error type and frequency expressed as a percentage of total errors

Error Type	Frequency	%
[Item] to [Incorrect place]		
Introduction to Method	1	1.69
Introduction to Results	1	1.69
Introduction to Discussion	16	27.12
Method to Introduction	3	5.08
Method to Results	5	8.48
Method to Discussion	0	0.0
Results to Introduction	0	0.0
Results to Method	5	8.48
Results to Discussion	12	20.34
Discussion to Introduction	7	11.87
Discussion to Method	3	5.08
Discussion to Results	6	10.17

As before the greatest difficulty readers had was distinguishing between the introduction and discussion sections, these accounting for almost 40 per cent of errors. Inability to distinguish between the results and discussion sections accounted for 30 per cent of errors while the method and results distinction proved the stumbling block in 17 per cent of cases.

7.6 General discussion

Readers experienced in the use of this text type seemed to possess a superstructure or model of it which enabled them to predict with high levels of accuracy where information was located. The existence of this superstructure probably results from the relatively standard form of such articles. There are few published accounts of experimental work in this (and other) disciplines that do not conform to this type. Obviously, frequent readers of this text type would acquire such an awareness of form over time which enabled them to predict likely location of information with minimum effort.

However, it is also worth noting that the classic IMRD structure acts as a framework for, or model of, the scientific process itself. Scientific research usually takes the form of examining the current literature to formulate a hypothesis for investigation, designing an experimental procedure to test this hypothesis, gathering and analysing data, and finally examining the results in the light of other work. Each of these activities has its parallel in the resulting description i.e. the experimental report. Generations of undergraduates are taught this model of investigation and reportage (even if it is, as Medawar (1964) stated, more a reflection of what scientists would like readers to think they have done rather than what they actually did!) so it is not surprising to find superstructures for this emerging. In a very real sense therefore, text structures can reflect conventions and standards of behaviour and cognition as argued by van Dijk (1980) and van Dijk and Kintsch (1983).

Regardless of any hypothetical cognitive representations underlying text usage, what is interesting from a human factors perspective is the high degree of accuracy shown by all readers in these experiments. From a rapid scan of the available text they can deduce the most likely location of that part in the whole and by extension, what is likely to precede, accompany and follow it. The results of experiment two clearly demonstrate that this representation holds also for screen presented text.

If designers are to consider seriously alternative structures for electronic or hypertext versions then they would need to overcome this acquired processing tendency of experienced readers. This is an all too unlikely occurrence given the embedded nature of this representational structure in the minds of readers, the teaching of the scientific process and the communication format of scientists.

Thus a hypertext journal article would need to retain the broad structure of the paper versions if it is to be immediately usable. For example, keeping the major headings and their standard order as the 'backbone' of the text would facilitate rapid exploration of required sections and narrow the search space for information location. Combined with the rapid access facilities of hypertext, such a format could result in the development of an electronic text that would ideally suit several of the reading tasks common to this text type identified in the journal usage study. This course of action was decided upon by the design team of the hypertext journal database discussed earlier. Rather than deconstructing the typical article format and linking information thematically, it was decided that the obvious presence of structural awareness in the readers of this text type justified its retention. Details of the resulting design can be found in McKnight *et al.* (1991).

7.7 Conclusions

It is perhaps to be expected that the format of academic articles is very familiar to experienced readers of this text type. However, the main point of these studies was not to confirm this fact but to examine the extent to which the perception of structure influenced readers organization of the text. The ease and speed with which these subjects arranged the material or predicted its location suggests that for this text type at least it is a very potent aid to organization. Other text types are likely to have less clear superstructures and in these cases, alternative structures for hypertext versions should be investigated. What seems likely though, is that readers do acquire some knowledge of structure for all texts, and that it increases with experience in using that text type. In use, it is likely to combine with spatial memory for layout (Rothkopf, 1971) to form a mental map of the text being read, facilitating searching and browsing of the material. Such issues must be addressed by the designer of any text presentation system if usability is to be ensured. Cognitive science provides the theoretical insight into the problem but designers will often need to resort to empirical methods to establish the nature of the structure best suited to any text type.

Notes

1. Though both systems are relatively recent (particularly ADONIS which has yet to go on full release) the time between specification and delivery of a system can be considerable. The original BLEND specifications were probably drawn up in or around 1980–82; the ADONIS workstation referred to in this chapter was formally specified in 1986 although discussions about its exact form and content commenced in 1980. The current database referred to here was specified in 1989 (though planning started in 1988). Such differences may appear short but are in fact a long time in information technology terms where the state of the art changes rapidly.

8

A framework for the design of electronic texts

It is a capital mistake to theorize before you have all the evidence.
It biases the judgement.

Sir Arthur Conan Doyle, *Sherlock Holmes: Study in Scarlet* (1904)

8.1 Theory versus empiricism: the strained role of human factors in the design process

Ergonomics is often criticized for being piecemeal rather than coherent, evaluative rather than predictive, and addressing specific issues in a way that leaves little scope for generalization of findings (Chapanis, 1988). This is reflected in the ergonomics practices carried out in systems design where human factors are often considered at a stage too late to effect better designs, i.e. the human factors specialist is seen as having a primary role in testing instantiated designs rather than influencing the initial specifications. Yet, given the much sought after opportunity to become involved in the early specification phase, ergonomists suffer from a lack of the conceptual tools and techniques necessary to overcome resistance from engineers seeking better inputs than vague or inflexible guidelines and exhortations to 'know the user'.

As outlined in Chapter 2, the standard philosophy underlying much human factors work is that of iterative user-centred design involving the development of prototypes and their subsequent evaluation, leading to further prototyping and so forth. While such an approach, properly executed, makes the development of usable technology more likely, it is a non-optimum process which can prove extremely expensive in terms of time and resources. Few design companies therefore are willing to invest the necessary effort to iterate through several cycles (Hannigan and Herring, 1986). This had led to the attempted reduction in the number of iterations needed and a move to bring human factors inputs into the design process earlier (Eason, 1988; Catterall *et al.*, 1989). Current emphasis is on rapid prototyping facilities which allow designers to mock-up disposable simulations quickly and cheaply. These can help, but even then the quality of the original prototype is dictated by the accuracy of the designer's conceptualization of the intended users, as outlined in Chapter 2. This is an area that requires much human factors work.

According to Card *et al.* (1983), what is needed to aid designers is an applied science of the user that is theory-based rather than empirical, using a common framework to provide insight and integration. By this they seem to mean that analytical techniques that do not require any empirical input could be used at the earliest stages of the process. As noted in Chapter 4, they propose a constrained version of cognitive psychology, with its emphasis on the information processing aspects of humans as the most suitable vehicle for this science and argue that if it is to have an impact, such a science must be based on task analysis, calculation and approximation which would lead to quantitative performance models of users. The role of such a framework would be to encapsulate some relevant knowledge of the user (often termed a 'user model', 'user typology' or 'user stereotype') and/or the task (similarly termed 'task model' etc. by some writers) that could provide guidance to the designer specifying the system. According to Card *et al.* the true role of an applied psychology is to provide such performance models in quantitative form for designers.

In reality, the apparent extremes of frequent empirical iterations and formal theory-led designs are merely opposite ends of a continuum; ends in which few human factors practitioners (as opposed to academics) permanently reside. More common is a mixed approach linking empiricism to theory and vice versa, with a bias towards empiricism due to the perceived lack of relevant (i.e. applicable) theoretical models at this time. The mixed approach is probably inevitable as both extremes are impossible to implement absolutely. All observation is theory impregnated according to most contemporary philosophers of science (e.g. Chalmers, 1976). Thus any artefact is going to be coloured by assumptions about the user, however implicit, when it is being developed prior to testing (and in philosophical terms one could argue that even the design of the experiment to test a system reflects underlying belief systems and is therefore theory impregnated). The empirical route to design can therefore in no way claim to reject totally theoretical perspectives of the user in favour of experimental facts. However, complete theories of human performance in HCI (or anywhere else for that matter) are non-existent and any design based on theoretical models alone must be evaluated by empirical means to ascertain its true level of usability.

Therefore, a practical goal for frameworks and models in HCI is to guide the derivation of suitable initial designs which, by virtue of their accuracy or utility (and these terms are not equivalent), reduce the number of iterations required before an ultimately acceptable design is achieved. Evaluations would subsequently act as confirmation or rejection of the design (or parts thereof) and if the latter, lead to refinement of both the resulting system and the theoretical framework underpinning it. The value of frameworks or models therefore lies both in their ability to reduce iterations and to be modified, if necessary, in the light of data from users and subsequently applied to other designs.

8.2 Frameworks and models: a clarification of terminology

The terms 'framework' and 'model' tend to be used interchangeably in the HCI literature, with model being the dominant descriptive term for such theoretical views of users. However, for purposes of clarity, the present work will draw a distinction between them in the manner described previously by Whitefield (1989). He describes a framework as a generic representation of the important aspects of the user and a model as a specific representation of those aspects in relation to a task. In this sense a framework provides the perspective of the user (i.e. reader) for all instances of interest while the model is derived according to the interaction of particular task demands and user. In these terms, the GOMS approach of Card *et al.* (1983) may be described as a framework from which specific performance models are derived. Whitefield (1989) uses the term framework to apply to the 'blackboard' architecture of AI theorists as in the speech recognition model of Hayes-Roth (1983), which he borrows to model specific instances of problem-solving behaviour in engineering design. In the present case a framework is proposed of the generic aspects of the reader which it is hoped will support the derivation of more specific models of reader–text interaction for particular tasks.

Several important criteria impinge on any proposed framework beyond the obvious one of utility. First, it must be accurate. This is not to say that it must offer a precise picture of the user or reader and text interaction being supported but what it offers should be correct in the sense that it describes real factors or aspects that influence the reading process. Second, it must be relatively non-complex. Invoking psychological descriptors or cognitive structures in a form suitable for non-specialists to use and apply is a difficult but necessary part of a good framework. Third, it must be suitably generic to be of relevance to more than one application. Just as the reading process covers myriad texts and tasks, designers should be able to utilize the framework describing this process for guidance on the design of more than one text system. Finally it should be modifiable. This does not mean that it must be altered every time it is used but that it should be capable of being adjusted in the light of feedback. The following descriptive framework is an attempt to satisfy all four criteria. The next section outlines the framework in detail.

8.3 The proposed framework

The framework is intended to be an approximate representation of the human cognitions and behaviours central to the reading process that are employed in the interaction between reader and document. It consists of four interactive elements that reflect the primary components of the reading situation at different phases. These elements represent the major factors deemed to be important to reading on the basis of the work in the preceding sections. They are:

1. a task model (TM) that deals with the reader's needs and uses for the material;
2. an information model (IM) that provides a model of the information space;
3. a set of manipulation skills and facilities (MSF) that support physical use of the material; and
4. a serial reading processor (SRP) that represents the cognitive and perceptual processing involved in reading words and sentences.

These are not isolated variables but interrelated components reflecting the cognitive, perceptual and psychomotor aspects of reading. In other words, according to this framework, reading is not a matter of merely scanning words on a page or acquiring and/or applying a representational model of the text's structure but a product of both these activities in conjunction with manipulating the document or information space and defining and achieving goals (all within a certain context). So a reader recognizes an information need, formulates a method of resolving this need, samples the document or information space appropriately applying their model of its structure, manipulates it physically as required and then perceives (in the experimental psychological sense) words on text until the necessary information is obtained. Obviously this is a very simple picture of the reading process; other more complex scenarios are possible such as the revision of one's reading goal in the light of new information or modifying one's initial information models to take account of new details and so forth. The point here is that regardless of the precise scenario, the elements described here should cover all important aspects from the point of view of the text designer. They are the building blocks of the activity described as reading which can be combined in numerous permutations. Each of these elements and their various interactions are described in more detail in the following sections.

8.3.1 The Task Model (TM)

The notion of the reading task as the crucial factor in understanding text use provides a sound basis from which electronic text design can be investigated. Readers interact with texts purposively, to obtain information, to understand, to learn etc. To do this they must allocate cognitive resources to some form of task model that decides what it is they want to get out of the text and also reviews their progress and, if necessary, revises the task.

This notion of intentionality in reading gives rise to the idea of planning in the reader's mind. The extent to which such plans exist is theoretically debatable but it seems reasonable to infer that the extent to which reading is a goal-driven behaviour some level of planning how to interact with the information source occurs.

From the task analyses and repertory grid studies carried out earlier it seems that such planning is relatively gross, taking the form of such intentions as 'go to the index, look for a relevant item and enter the text to locate the answer to my query' or 'to find out what statistical tests were used go to the results section and look for a specific description'. However, they can be much vaguer than these two examples which probably represent highly specified plans of interaction with the text. Reading an academic article to comprehend the full contents seems to be much less specifiable, the reader is more likely to formulate a plan such as 'read it from the start to the finish, skip any irrelevant or trivial bits, and if it gets too difficult jump on or leave it'. Furthermore, such a plan may be modified as the reading task develops e.g. the reader may decide that they need to reread a section several times, or may decide that they can comprehend it only by not reading it all. In this sense planning becomes more situated (e.g. Suchman, 1988) where the reader's plans are shaped by the context of the ongoing action and are not fully specifiable in advance.

Whatever the precise nature of the plan it seems appropriate to posit a task model that generates some method for dealing with the document or documents under consideration. Thus the framework must accommodate such activity in order to focus designers' attention on such reader-text interaction characteristics.

8.3.2 The Information Model (IM)

Readers possess (from experience), acquire (while using) and utilize a representation of the document's structure that may be termed a mental model of the text or information space. Such models allow readers to identify likely locations for information within the document, to predict the typical contents of a document, to know the level of detail likely to be found and to appreciate the similarities between documents etc. The journal and manual usage studies as well as the experiments in the previous chapter highlighted the existence of such knowledge and it is postulated here that such representations exist for all commonly used text types.

However, as outlined in the previous chapter, it is worth making a distinction here between 'global' and 'instantiated' schemata with regards such mental models (Brewer, 1987). In the present context a global schema consists of a representation of how a typical text type is organized, e.g. an experimental article is typically made up of introduction, method, results and discussion sections, or a newspaper is made up of a series of articles covering a range of topics grouped into sections on politics, sport, finance etc. These are the type of structural representations that are general and exist independently of any specific document (though of course they only emerge over time after frequent interactions with many documents).

An instantiated schema consists of an embodiment of the generic model based on exposure to a specific text, e.g. noting that the particular article one is reading has a very short introduction or there is a reference to 'Bloggs (1982)' on the top of a right hand page containing a figure. In other words, when a reader interacts with a text, the original structural model of the text type becomes fleshed out with specific details of the particular text being read. The terms 'global' and 'instantiated schema' are overly technical however, therefore this distinction will be referred to here more simply as the difference between a model (which is generic) and a map (which is specific). In these terms readers can be said to form mental maps of particular texts as they use them, models help them in this but are not themselves essential for map formation (i.e. it is assumed that a reader can form a detailed map of a document without having been exposed to similar types of text before). In this way, frequent map formation with a document type can be seen as supporting model formation of that document type's generic structure.

In use, the information model helps the reader to organize the text's contents by fitting it into a meaningful structure and thus guards against navigational difficulties by providing context, i.e. it supports the formation of a mental map of the information space. Thus what is initially a model becomes, with use, a map of a specific text. Where no model exists in advance, a map can be formed directly. The point at which a model becomes a map is difficult to quantify and probably not pertinent to present needs. After all, knowing one structural detail about a specific text hardly conveys the idea of a map as the term is typically understood. Therefore the term information model is retained for general use in the framework, map will only be employed when discussing readers' detailed knowledge of a specific text's structure.

8.3.3 Manipulation Skills and Facilities (MSF)

Readers must be able to manipulate text. This simple statement hides innumerable complex issues in the design of electronic text. Except for very short passages, documents cover more than one page or screen and the reader needs to be able to physically alter their view of the material. Even in such situations they still need to locate and 'open' the text for reading, actions which clearly involve manipulation. With paper, such skills are acquired early by readers and are largely transferable from one text form to another. If one can manipulate a paperback novel one will have few difficulties with a textbook and so forth, although there are obvious exceptions in the paper domain and the ability to easily manipulate broadsheet newspapers in confined spaces is a specific skill that is relatively unique to that text form. Anybody who has witnessed the elegant manipulations of individuals reading broadsheet newspapers such as *The Times* or *The Guardian* while travelling in the crowded carriages of the London Underground will appreciate the skill factor involved.

However, such paper-based skills are limited in terms of what you can do with the text. Most readers are skilled in using their fingers to keep pages of interest available while searching elsewhere in the document or flicking through pages of text at just the right speed to scan for a particular section, but beyond these actions, manipulation of documents becomes difficult. When one then considers manipulation of multiple documents these limitations are exacerbated.

Large electronic texts are awkward to manipulate by means of scrolling or paging alone but the advent of hypertext with its associated 'point and click' facilities and graphical user interface qualities has eased this somewhat. However, the immediacy of interaction with electronic text is less than it is with paper by virtue of the microprocessor interface between reader and information on screens. Furthermore, the lack of standards in current electronic information systems means that acquiring the skills to manipulate documents on one system will not necessarily be of any use for manipulating texts on another. Obviously electronic text systems afford more sophisticated manipulations, such as searching, which can prove particularly useful for certain tasks and render otherwise daunting tasks (such as locating every reference to a certain topic in the complete works of Shakespeare) now manageable in minutes rather than days. Yet such facilities are not always a guarantee of accurate performance.

The various advantages and disadvantages of manipulation facilities on screens have been presented in detail in Chapter 3. Ultimately, the goal is to design transparent manipulation facilities that free the reader's processing capacity for task completion. Slow or awkward manipulations are certain to prove disruptive to the reading process. The framework raises these issues as essential parts of the reading process and therefore important ones for designers to consider in the development of electronic text.

8.3.4 Serial Reading Processor (SRP)

The final element of the framework is the serial reading processor. It is proposed that this is the module that actually perceives the images from the document and carries out the activities most typically described as 'reading' in the psychological literature (e.g. Just and Carpenter, 1980). Thus eye movements, fixations, letter/word recognition and other perceptual, linguistic and (low-level) cognitive functions involved in extracting meaning from the textual image are properly located at this level.

The prefix 'serial' on the reading processor emphasizes the perspective that at this level of cognition reading generally occurs in a serial fashion. That is not to say that texts are accessed and used serially (which, despite the many claims of the technocrats, they are obviously not according to the evidence reviewed in this book) but that at the level of engagement between the eyes and the sentence reading is, for the most part, serial. Obviously regressions occur and people jump about from one part of a page to another without physically manipulating the document but even then, this requires readers' attention to focus briefly on their model of the text and so changes the processes that are being dealt with. At the level detailed here, information extraction from a document relies on the reader serially processing letters, words and sentences.

The question of how serial reading is accomplished will not be discussed further. Decades of psychological investigation have been spent looking at the question of how humans read and some of the conclusions drawn from this work have been discussed in Chapters 3 and 4. Present emphasis dictates that the findings on eye movements, reading speeds, letter and word recognition etc. are considered sound but are of relevance here only to the extent that reading electronic text is influenced by or alters these aspects of the process. An obvious example of how issues at this level affect electronic text design is to be found in the image quality work of Gould *et al.* (1987a; 1987b).

8.4 Interactions between the elements

So far, the basic components of the framework have been described. These reflect the human aspects of performance during the reading process and are therefore the elements that seem pertinent to electronic design. A schematic representation of the framework is presented in Figure 8.1. As shown, the elements are all related and collectively framed within the context in which the activity occurs.

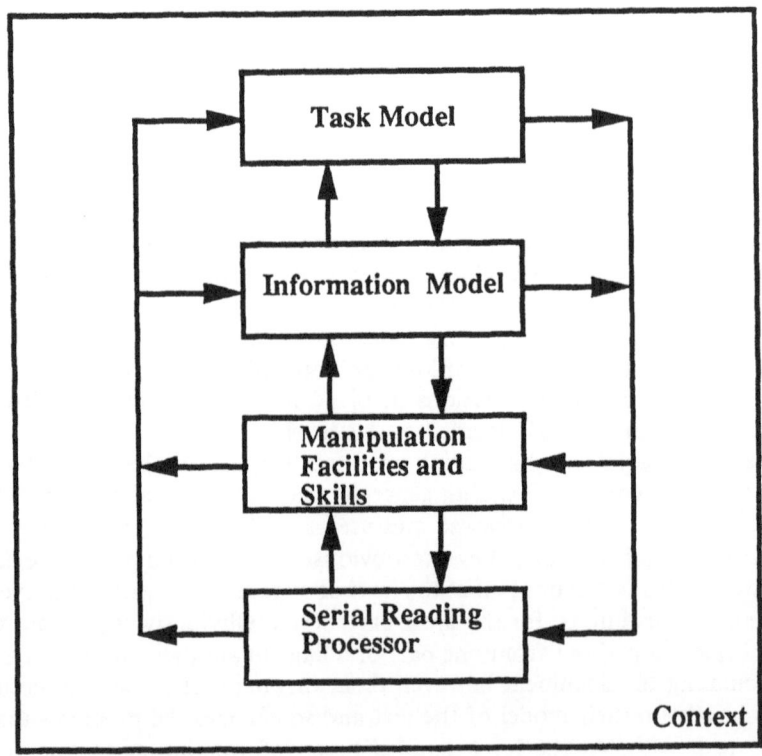

Figure 8.1   The framework for describing reading.

According to the framework there are 12 possible interactions between these elements that can occur. These will be described individually.

Task model to information model

When a task is formulated the reader usually interprets or mediates its formulation and expectation of its outcome in terms of their model of the information space. For example, if the task is 'Find the reference in the text to Bloggs', the reader applies the model to narrow the search space and produce an inference such as 'It is more likely to be in the section labelled "Related Work" than in "Results"'. This is a routine and rapid occurrence.

Task model to manipulation skills and facilities

Where an information model does not exist (when this is a reader's first exposure to a text type for example) the reader, upon formulating a task may proceed to manipulate the text without any knowledge of its layout or contents. This may be as simple as opening a book (or electronic file) with no better intention than reading it from start to finish until the target information is located. In terms of the framework, this is a case of direct interaction between the TM and the MSF elements. It implies that absence of an information model at the outset does not prevent text usage.

Task model to serial reading processor

In cases where the text is short and available, e.g. a single page memo on one's desktop, even manipulation facilities may be unnecessary. Similarly during a particular sub-task of a larger one, a reader may engage only TM and SRP to perform that sub-task, e.g. locate a word in a paragraph currently on screen or on the open page. This example highlights the fact that it is possible to perform some reading acts without engaging either in IM or MSF elements.

Information model to task model

A model of the information one is dealing with can influence the type of task one tries to perform with it and aid accurate specification of that task. For example, if a reader's goal is to find out about particular theories of child development, a model of the text (e.g. an introductory psychology textbook) could suggest that the book was inappropriate but it might offer suggestions for further reading. On the other hand, such an interaction between IM and TM elements could occur where, after reading the text for several minutes the reader's evolving model might indicate that the text is unlikely to contain the form of information required and therefore the task needs to be re-specified.

Information model to manipulation skills and facilities

The interaction between these two elements in this direction is likely to be of the form: model directing manipulation, e.g. the information being sought is at the end of the text therefore page or scroll to the last chapter. Such rapid interactions should characterize many reading situations.

Information model to serial reading processor

Again, this is only likely to occur for particular tasks and very short texts. An example might be identifying the sender of a one page letter. In this case one's model of the letter form suggests that an address may be provided at the top of the page or a signature will be present at the bottom. Once the letter has been opened and the page unfolded further manipulation activities can often be bypassed.

Manipulation skills and facilities to serial reading processor

Once the task and model aspects have been applied and the text is of the type that will require manipulation, an interaction between the MSF and SRP elements occurs. A simple example is the reader turning a page to allow reading to commence.

Manipulation skills and facilities to information model

Such an interaction might be expected to result when, faced with an unfamiliar text, the reader manipulates it and induces the formation of a primitive information model (the 'flick through to see what's in it' approach).

Manipulation facilities to task model

In this instance the information flows directly from the element concerned with text manipulation to the task model. Though presumably rare, an example might be when a reader finds that they cannot search for a term and therefore cannot perform the task as originally envisaged.

Serial reading processor to manipulation skills and facilities

When the reading processor is interrupted by a page break or screen end an interaction with manipulation facilities occurs to facilitate further SRP activity with the text. This is logically distinct from the MSF to SRP activity described earlier which refers to activity occurring prior to SRP activity.

Serial reading processor to information model

The reading processor may interact directly with the information model by providing information to the reader about the contents of a page or the type of material contained in the document at that point, thereby supporting the formation of a map, for example, noting the occurrence of a particular word or phrase as a potential landmark in the document.

Serial reading processor to task model

The direct interaction between the two extreme elements in the model may occur in this direction when, for example, information is read which solves the immediate task or sub-task, e.g. if, when searching for a word or phrase in a certain section, the reader perceives it and thereby resolves the immediate task without requiring further manipulation or model activities.

In practice, it is not hypothesized that such neat interactions occur in isolated units. Meaningful engagement with a document is more likely to result in multiple rapid interactions between these various elements. For example, a scenario can be envisaged where, reading an academic article for comprehension, the task model interacts with the model to identify the best plan for achieving completion. This could involve several TM→IM and IM→TM interactions before deciding perhaps to serially read the text from start to finish. If this plan is followed then manipulation facilities come into play and serial reading commences. The MLS→SRP interaction and SRP→MLS interaction may occur iteratively (with occasional SRP→IM interactions as distinguishing features are noted) until the last page is reached at which point attention passes back ultimately to the TM to consider what to do next.

Also, the speed and the iterative nature of the interaction between these elements is likely to be such that it is difficult to demonstrate empirically the direction of the information flow. In many instances it would be virtually impossible to prove that information went from MSF to IM rather than the other way and so forth. However this does not preclude examination of these elements and their interactions in an attempt to understand better the process of reading from a human factors perspective. The elements reflect the major components of reading that emerged as important from the studies earlier in this book and are intended as a broad representation of what ergonomically occurs during the reading process.

8.5 *Qualitative versus quantitative representations*

The framework presented here is a relatively simple representation of those issues found to be of importance to the usability of an electronic text. They are described in this framework qualitatively, i.e. their actions are presented in general terms rather than being specified formally. The absence of rules of exclusion/inclusion or numerical values necessitates greater interpretation by an eventual user of this framework than a quantitative framework or model of HCI or reading. This is intentional, a matter of choice (and some necessity given current knowledge) rather than a failure on the author's part to specify further the framework's components. In the present section the case for such a framework is presented by considering it in the light of the usage of typical quantitative models more commonly expounded in this domain.

As outlined earlier in the book, traditional psychological models of the reading process are often very detailed, postulate the existence of numerous cognitive structures and processes, and tend to concentrate on isolated or limited aspects of the reading process such as word recognition, sentence processing or eye movements. It has been argued at length in Chapter 4 that the level of detail provided by such models of human information processing is too low to be applied in HCI and that in the case of reading, this severely hampers the development of usable electronic text systems. The form of modelling common to cognitive psychology is mirrored sharply in the attempts of human factors professionals to describe or model human behaviour at the computer interface. As detailed in Chapter 4, the major research effort has concentrated on developing formal models of a quantitative kind for designers to apply at the specification stage of product design.

Advocates of the quantitative approach cite precision, non-ambiguity of terminology and ability to calculate design trade-offs as major advantages of such models (e.g. Harrison and Thimbleby, 1990). While this may be true for models such as GOMS or Cognitive Complexity Theory (CCT) (Kieras and Polson, 1985) when used for very specific analyses (and there is little by way of confirmatory evidence of this yet) there are two underlying assumptions in this view which are directly pertinent to the present work. One is that designers find such quantifiable outputs relevant and the other is that the human performance and behaviour one is interested in can be reduced to suitable numeric functions. There are many further criticisms of these formal methods that could be made which are not directly pertinent to this discussion e.g. what of their accuracy? why do independent users of the methods often derive different models of the same problem? or why are they so difficult to use? Critiques of these aspects can be found in Carroll and Campbell (1986), Winograd and Flores (1988) and Sharratt (1987).

From what is known about the way designers work in realtime, theoretical quantitative models seem to have little relevance in their current form (Buckley, 1989; Carroll, 1990). Virtually all successful reports of the application of these models emerge from experimental work in academic rather than industrial environments (e.g. Polson *et al.*, 1986, 1987). Their proponents might claim that they are useful and reliable but the design community remains unconvinced. This could result from several reasons not related to the scientific validity of the approach such as the difficulty of applying them (they usually require substantial domain knowledge to be used effectively) or their concern with narrow aspects of tasks rather than global user behaviour, which renders them more suitable for application after the initial specification rather than before. According to Landauer (1987) such models do not tell a designer how to design a good system in the first place (which is what designers really want to know). Instead, they just advance the moment when evaluation can first be carried out to the pre-prototype stage, i.e. they are a measurement tool rather than a creative design aid.

The present framework takes such shortcomings as its starting point and is intended to offer a conceptual aid to electronic text designers that addresses such problems. First, a designer does not require sophisticated knowledge of human cognition or the psychology of reading to comprehend the framework. Obviously detailed psychological work underlies concepts such as information models, task models, etc. but the designer can consider the basic issues without possessing such knowledge. The more knowledge of human cognition that designers possess the more critically and usefully they may be able to apply this work, but such knowledge is not a prerequisite for use. Second, unlike most formal methods, the present framework does not require the use of a formal language or sequences of rules to support interpretations of likely user behaviour. It is intended only to draw attention to issues such as image quality and information organization in the first instance so that the designer realizes what is important in a design, not to provide a means of calculating design trade-offs in terms of performance times. Third, the framework covers the full range of behaviour described as reading as it impacts on system design, not just a particular subset of it. It is intended to cover reading as it pertains both to proof-reading and scanning of lengthy texts, or to using textbooks or magazines. Finally, it is suggested that a successful electronic text system is one that addresses all four elements of the framework in its design, therefore a designer can employ it to guide the initial specifications, i.e. it is a design aid more than a measurement tool.

Such explicit qualitative models of human performance are not commonplace in HCI (though implicit ones abound) but the framework is not without precedent. In psychological terms for example, guidelines as simple as 'the human is an information processor with five sense channels' may be considered a highly simplified qualitative model. This is a particularly broad representation and as such, is of very little use to designers. Norman (1986) provides a more specific model of human interaction with systems which consists of seven stages ranging from forming an intention to act to evaluating an outcome in terms of goals sought. Dillon (1987) presents a qualitative model of user knowledge development in terms of three stages: confusion, rationality and knowledge. Neither of these offer quantitative power to the designer but they do elaborate or make explicit certain 'myths' or standard beliefs about users, e.g. that they are goal seeking (Norman's model) or they get better in a predictable stage-wise sequence with experience (Dillon's model). They do this in a way that their proponents hope will both help designers and reflect psychological reality.

A suspicion exists that qualitative approaches are inherently vague, are more likely to be rejected by engineers (who supposedly prefer numbers to words) and, in the light of the aggressive 'selling' in the human factors literature of the hard quantitative approaches, are somehow less scientific however that is measured. This need not be so however. The literature on design (not specifically HCI-related) outlined briefly in Chapter 2, has clearly demonstrated that designers tend to rely heavily on heuristics, intuition, and 'try it and see' approaches rather than the rigid hypothetico-deductive logic based approaches manifest in trained scientists. Qualitative models could well offer the form of guidance more suited to this type of problem solving than some time-consuming but powerful quantitative approach.

As stated earlier, implicit qualitative models abound. All designers and ergonomists, in fact everyone involved in the development of a product, from the marketing department to the specification writers, have an implicit model of the users and tasks that the end product will support. These models just vary in detail and accuracy depending on the possessor's role. The sales representative presumably has a view of the user as a customer while the marketing person might view users as belonging to certain job, skill and economic categories. These views or representations of the target users are 'models' as such.

For present purposes, the main interest is in the models possessed by the designers and ergonomists. The latter participant, by virtue of probable training in a human science such as psychology, is likely to model the user as an information processor with cognitive dispositions, skills, habits and preferences. This model will probably include detailed knowledge of cognitive components such as short-term memory, long-term memory, mental models etc. and their potential impact on the usability of a computer system. On the basis of task analysis and previous experience, skilled ergonomists can derive a set of user characteristics for input to design specifications. In a very real sense then, this is a form of qualitative modelling. By extension, quantitative model proposers must have their own qualitative models on which they base their formalisms.

Designers, as described earlier, will always implicitly model the user in drawing up specifications. However, their models of the user and task tend to be ill-formed and vague, based often more on intuition rather than facts. Yet the resultant usability of a product is largely determined by the quality of the implicit model underlying the design and mistakes made at this point in the product life cycle are considered to be the most expensive to rectify (Dunn, 1984). A progressive aim of human factors inputs therefore must be to improve this model, either by quantitative or qualitative means. The evidence on balance would suggest that a suitable qualitative model is likely to be more relevant to designers than existing formal quantitative ones.

What needs to be improved is the explication of these models. Vague descriptions of user characteristics are probably better than nothing but guidelines and handbooks of design principles are rarely successful. Opting to present a more structured view in terms of a framework describing the relevant components of the user-system interaction, as embodied in the present framework, is likely to have more relevance.

Rasmussen (1986) advocates the use of qualitative models in this sense. He argues that quantitative modelling concentrates on one level of behaviour, particularly sensori-motor in well-practised tasks, which is inappropriate for the type of higher level cognitive functioning of interest to many designers. For him, the major distinction between the two forms of model is not that one is respectable or scientific and the other intrinsically soft and vague, but that the qualitative type concentrate on broad categories of behaviour while the quantitative focus on specifics. He rejects the traditional argument that the former are merely undeveloped or premature quantitative models and states that designers of computer systems might well find qualitative models of direct relevance to their work in the design of any system where users have some choice on how they will work.

This is an important but often overlooked difference between current and past technologies. Interactive computer systems afford greater user control than traditional mechanically engineered machines which had to be operated in a set manner. Interesting conclusions could be drawn from this on a range of issues from the changing nature of work to the more enlightened socio-political views of workers in contemporary organizations. Whether such conclusions would be valid however, is another issue (see e.g. Eason, 1988).

In the case of electronic text design it has been strenuously argued that the quantitative approach is not appropriate. Whether this is a function of current knowledge limitations or inherent failings in the approach is not of direct concern in this book, but philosophically at least, the present author's inclinations are in the latter direction. The type of knowledge needed by the designers at HUSAT is often of the generic qualitative kind. Furthermore, the act of reading, as it is interpreted here involves behaviours and cognitions too broad to fit the 10-second boundary of the classic GOMS approach. Therefore, regardless of the ultimate success of a quantitative analysis of cognition, qualitative models do seem to have relevance at this time and are worth pursuing as design aids in HCI.

8.6 The utility of the proposed framework

What use is this framework to designers of electronic text systems? It is possible at this time to specify three potential uses. In the first instance the framework is useful as a guiding principle or type of advance organizer of information (Ausubel, 1968) that gives the designer an orientation towards design enabling them to bring relevant knowledge to bear on the problem.

Secondly, by parsing the issues into elements, the framework facilitates identification of the important ones to address. This framework suggests four levels of issue to consider in any context: the user's task and their perception of it; the information model they possess or must acquire; the manipulation facilities they require; and the actual 'eye-on-text' aspects involved.

In the third instance the framework provides a means for ensuring that all issues relevant to the design of electronic text are considered. It is not enough that research is carried out on text navigation and developers ignore image quality or input devices (and vice versa). A good electronic text system will address all issues (indeed it is almost a definition of a good electronic text that it does so).

The above applications consider the uses directly to designers at the first stages of system development. In this sense the term designer encompasses any ergonomists or human factors professionals seeking to influence the specification of an application. However, the framework also has relevance to later stages of the design process such as evaluation. In such a situation the framework user could assess a system in terms of the four elements and identify potential weaknesses in a design. This would be a typical use for expert evaluation, perhaps the most common evaluation technique in HCI.

Outside the specific life cycle of a product, the framework has potential uses by human factors researchers (or professionals less interested in specific design problems) in that it could be used as a basis for studying reader behaviour and performance. The framework is intended to be a synopsis of the relevant issues in the reading process as identified in the earlier studies. Therefore, it should offer ergonomists or psychologists interested in reader–system interaction a means of interpreting the ever-expanding literature in a reader-relevant light. These issues will be discussed further in the next chapter.

9

Assessing the framework in terms of validity and utility

You can't learn too soon that the most useful thing about a principle is that it can always be sacrificed to expediency

W. S. Maugham, *Circle* (1921)

9.1 Introduction

The framework as described in the previous chapter derives from the various analyses on readers' classifications of texts, descriptions of their usage and the experimental investigation of their models for one text type. While it represents an intuitively coherent categorization of the issues involved in the reading process, it cannot at this stage lay claim to anything more. The obvious questions to ask now are: is this a valid way of considering behaviour? and what purpose does such a framework serve? The present chapter concentrates on answering these questions.

9.1.1 The validity issue: is this a fair description of the reading process?

In theoretical terms, validity refers to the extent to which any psychological concept or model can be viewed as an accurate representation of that which it purports to describe. It is an important issue, e.g. for test developers in psychometrics who have devised appropriate means for calculating validity coefficients, i.e. ratings of the extent to which a test really measures what it claims to, be it verbal intelligence, personality or whatever (e.g. Anastasi, 1990). This is usually achieved by reference to an external objective criterion, e.g. scores on a test of aptitude for medicine could be compared with subsequent performance at medical school – a test with high validity should provide scores which correlate significantly with performance at college. For many other psychological constructs such as those typically postulated by cognitive psychologists, e.g. short-term memory buffers, mental models, etc. validity is far less amenable to assessment.

In many ways, one can only test the validity of such concepts by trying to prove their invalidity, an odd description perhaps of a process known to scientists as falsification (Popper, 1972). That is, the construct (or theory, concept, model, etc.) leads to an experimental hypothesis, which the researcher tests by empirical means. If the hypothesis is not supported, the construct or theory under test is more or less falsified and therefore undergoes modification before the whole process starts again. If the hypothesis is supported, then the theoretical structure is not considered true, but is subjected to further tests. Even if a theory is not falsified after numerous tests it is never considered true, merely adequate, given current knowledge. This is a somewhat idealized (and Popperian) view of science; in truth, many scientists are loath to subject pet theories to rigid cyclical scrutiny and the scientific process shuffles between contrasting views, numerous modified theories and occasional revolutions (e.g. Kuhn, 1962).

The relevance of these issues to the present case is that if a test of predictions made on the basis of the framework proved positive, this would not imply that the framework was valid psychologically, only that it had not been falsified. The best that can be done is to try and match data from readers to the framework and identify the extent to which the framework would explain these. Any mismatch would be deemed as falsifying the framework as it currently stands and necessitate its modification.

This process is more difficult than it seems however. Obtaining the type of data that would truly test the existence of such elements as task models or serial reading processors is riddled with philosophical problems, not least of which is what constitutes acceptable evidence? In contemporary cognitive science the development of a working software program that mimics the performance under consideration is seen as the ultimate test (e.g. Johnson-Laird, 1983). Failure to provide such support is considered a flaw in any proposed model. Thus, according to hardliners, if a hypothesised psychological process cannot be specified as an effective procedure it cannot be considered valid.[1]

This is a strong test for any psychological theory and, not surprisingly, few contemporary theories pass.[2] Such a test could be considered as demonstrating the invalidity of the framework specified in the previous chapter. However, this is not as fatal as it sounds. The framework is not intended to provide a precise model of human mental activity during reading. To test for this would therefore be pointless. In its form as a generic description of the reading process at a level appropriate for design however, it is proposed as valid, and a test of this would be relevant. One test of suitable form would be to examine readers' behaviour and verbal protocols when using a document, parse them into their various components and then relate these to the components in the framework. If the framework is valid, such protocols should provide clear examples of the behavioural and cognitive elements that constitute the framework. If it is an invalid description, the protocols should fail to provide such a match or should indicate the presence of activities not accounted for in the elements of the framework. It is this form of test that will be reported here.

9.1.2 The utility issue: is the framework any use in design?

The term 'utility' means relevance, pertinence or usefulness. Accordingly, in this context it implies any appropriate use that may be made of the framework in the electronic text design process. A test of this framework's utility could be made in several ways, ranging from the ideal to the feasible. In this section the potential test scenarios are examined for suitability in the present circumstances.

First, the framework could be presented to designers at the outset of a new design project, under varying degrees of control and manipulation, and its effects on several end products assessed. This would be an ideal test in that it would involve a controlled sample of designers applying the framework in similar task environments. However, under the dual constraints of the commercial pressures in software houses and the limited influence of a single researcher, such a test scenario is unlikely to be practicable. Not only would it require the type of interference in the normal design process that many companies would, quite understandably, actively seek to avoid, but the complex analysis that would be required to untangle the dependent variables in such a scenario would be extremely taxing and potentially beyond the abilities of any one researcher.

A near-ideal test might be to present it to a sample of designers working on a range of application development projects and ask for feedback from them on its utility at some specified future date, by which time they would hopefully have had the opportunity to apply it. While this might seem more feasible than the ideal approach, it would still require a commitment to the use of the framework in a commercial enterprise that would seem difficult to justify given its present form and status. Furthermore, it would require both the designers to use the framework as and when they saw fit, and the researcher to rely on sporadic feedback as the sole measure of utility. Given the experiences of many human factors professionals who have handed over design tools in this manner, only to come back later to find them gathering dust on a shelf, and the expected drop out in participation for any data-gathering exercise necessitating subjects to initiate their own responses, a high price would almost certainly be paid for a less than perfect study.

With these issues in mind a more suitable test of its utility in the present context would be to apply the framework to an experimental investigation of people reading texts and use it to predict the likely effects of design variables on performance. If the framework is to have any utility to designers it should at least be able to predict gross usage characteristics that are affected by presentation variables. This is a more manageable test of the framework in that it can be carried out without involving a commercial software house and their designers. Furthermore it is the type of test that must be seen as a prerequisite to any real-world testing. Such a utility test is described here.

9.1.3 The present situation

In the present chapter, two experimental studies are reported. In the first, an experiment carried out by the author in conjunction with two colleagues at HUSAT offered one suitable test vehicle for the framework. This was an investigation of readers' ability to extract relevant information from a text presented in four different formats: one paper and three electronic versions. The data presented here contain part of the original analysis carried out by the design team of which the author was a member (McKnight *et al.*, 1990a) plus a substantial analysis by the present author alone of data not used by the team in its original work. Though primarily concerned with the validity issue, this experiment also offers some insight into the utility of the framework.

A second experiment designed to test a specific prediction derived from the framework was also carried out by the author. This can be seen as a straight utility test in that it represents an attempt to employ the framework to guide predictions of user performance with a text presented on paper or screen. For convenience only (as the descriptors are not precise), the two experiments will be referred to in this chapter as the validity and the utility experiments respectively.

9.2 The validity experiment

This study examined readers' performance in extracting answers to questions from a short text on the subject of wine making. The envisaged task scenario was one where an individual, armed with the document, staffed an enquiry service where people would ask relatively straightforward questions on such topics as the largest winemaking regions of France or the meaning of terms such as 'second fermentation' and so forth. No previous knowledge of wine was required as the answers to all the questions were available in the text.

The aim of the experiment was to examine the extent to which two popular hypertext environments would support such a task compared with paper or a linear electronic text. To this extent it was a very open ended study concerned with exploring the issues rather than manipulating small independent variables. For present purposes the analysis will focus on the concurrent verbal protocols elicited from readers while also examining some of the performance data.

9.2.1 The application of the framework to the location task

The framework suggests that there are four major components to the reading task in any context, each of which is represented by a rectangular box in Figure 9.1. This schematic representation provides a descriptive model of the likely sequence of events involved in performing the experimental tasks.

Initially, it is suggested, the reader will employ task processing skills to formulate a means of resolving the task. It is probable that for the type of tasks involved in the present experiment, the reader will identify a search criterion from the question and attempt to obtain an answer by finding a relevant match to that criterion in the body of the text. For example, if the question is 'What types of wine are produced in the Loire region?', it is assumed that the reader is most likely to select 'Loire' as a target and locate

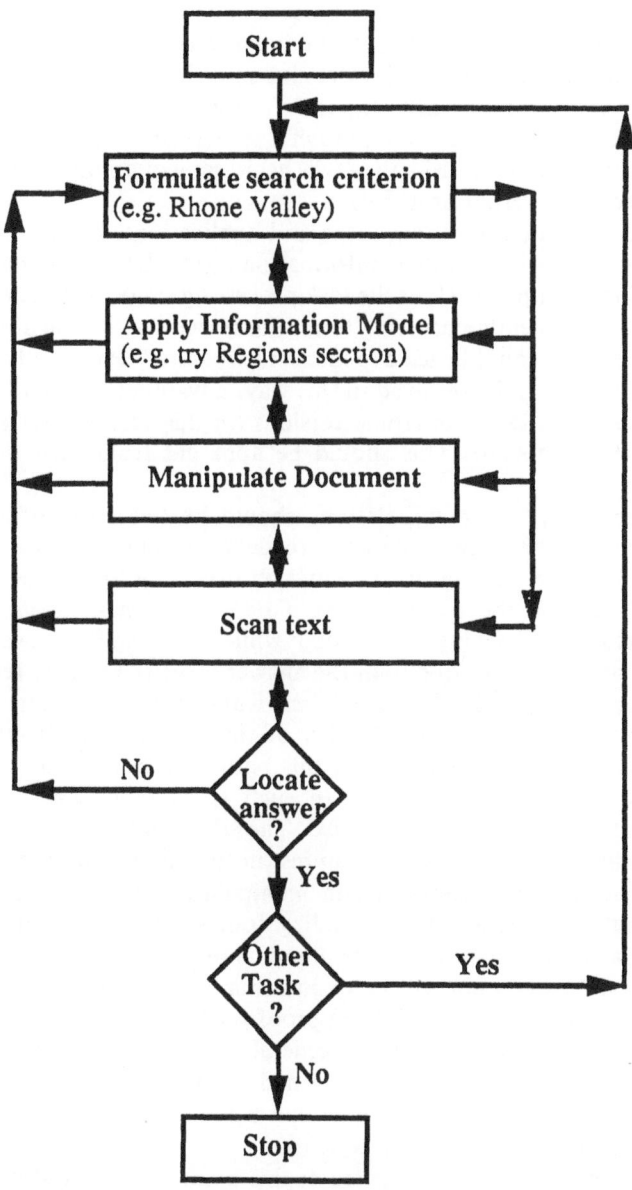

Figure 9.1 A schematic model of readers' behaviour on an information location task.

references to this in the text until a pertinent section on wine types is located. However, selection of any other target is possible and allowable within the framework.

Once a satisfactory search criterion is identified, according to the framework, the reader's attention switches to the information model which is used to guide the search for a matching term. It has been shown that the information model of certain text types is well-formed and supports such applications, but in the present context it is expected that the uniqueness of the text would be unlikely to afford a detailed model, particularly at the outset. However, even for unique texts, exposure to them facilitates the development of a map and it is likely that even though a reader lacks an information model at the start, after several tasks they will begin to acquire one. This should be apparent from the verbal protocols.

The reader could, for certain tasks and applications, bypass or overcome any inherent limitations in their information model by employing the search facilities of the computer. If, at the task processing stage, the reader deduces a specific and infrequently occurring term, the search facilities available on two of the applications could be used to locate the required text directly. Certainly, for the tasks that can be resolved in this way, advantages should be conveyed to readers using suitable electronic versions (in this case HyperCard and the word processor version). This should be apparent from their performance data.

For other tasks, the manipulations should be less straightforward. For example, one of the tasks required readers to compare two sections of information before gaining the information necessary to provide an answer. One could imagine the readers using a paper document opening the first relevant section and keeping a finger on it while searching for the second, prior to flipping between them to obtain the answer. This is a typical reader–paper text interaction but a difficult one to mimic without (and sometimes even with) windowing facilities on screen. Neither of the hypertext applications in this experiment supported windowing of this nature, although this was possible with the word processor version. In such tasks, one would hypothesize advantages to paper over electronic text. The advantages and disadvantages of manipulating text should also be manifest in the readers' protocols.

Once the reader has searched and manipulated the text, the framework suggests that a scanning type of reading follows. In other words, it is not expected that readers will read large amounts of text in a serial fashion while performing these tasks but will jump and skim read sections looking for cues and target words. From the work on proof-reading text on paper and screen reviewed earlier, it is clear that in general, the advantages lie with paper. Although few researchers have examined scanning as opposed to proof-reading, there is no reason to assume that a similar advantage to paper does not hold for this type of reading style too.

If the target is successfully located at this point then the task is completed and the reader can start on the next one. This initiates a sequence of events similar to those just outlined, though with each subsequent completion it is expected that the information model becomes more elaborate (i.e. a mental map of the document is being formed) and familiarity with the requisite manipulation skills grows. This should also be apparent in the protocols.

9.2.2 Readers, tasks and texts

Sixteen readers participated in the study, nine male and seven female, age range 21–36. All had experience of reading a variety of computer systems and applications. The text under consideration was a document titled 'Introduction to Wines' by Buie and Hassell (1982), a basic guide to the history, production and appreciation of wine. This document was widely distributed in hypertext form by Ben Shneiderman as a demonstration of the TIES (The Interactive Encyclopaedia System) package (now known as HyperTIES). In the TIES version, each topic was held as a separate file, resulting in 40 individual small files. For the HyperCard version, a topic card was created for each corresponding TIES file.

In its paper format, this text consists of 13 A4 pages of text with no figures and would thus be most aptly described as a booklet or essay type text. In order to place a structure on the document that would facilitate its presentation as a paper text the topics were retained in the linked groups of the hypertext original but ordered from start to finish in a manner that seemed intuitively sensible to the experimenters. Thus an introduction was followed by a general overview of the processes involved in manufacturing wine before specific countries and regions were presented. This structure was retained faithfully for the word processor version.

In order to test this intuitive arrangement for suitability a quick pilot test was carried out by the author. This involved asking readers to order a set of cards, each of which had a term on it referring to the title of each of the files in the TIES version. These were wine related terms such as 'Bordeaux', 'Production' or 'Aperitifs' rather like a list of contents. Three readers were each asked to group these into what they perceived to be a suitable single-document structure. The results confirmed the structure of the experimenters i.e. groups were formed out of countries and subordinate regions, wine manufacture, and particular wines and grapes.

The HyperCard and word processor versions were displayed on a monochrome Macintosh II screen and the TIES version was displayed on an IBM PC colour screen. The paper version was a card-covered spiral bound, A4 text.

Readers were required to use the text to answer a set of 12 questions. These were specially developed to ensure that a range of information retrieval strategies were employed to answer them and that the questions did not unduly favour any one medium. The answers to all questions were specifically mentioned in the text.

A four-condition, independent subjects design was employed with presentation format (HyperCard, TIES, paper and word processor) as the independent variable. The dependent variables were speed, accuracy, access strategy, readers' estimate of document size and verbal protocols.

Readers were tested individually in the usability laboratory at HUSAT. This consists of a simulated office environment containing video cameras and sound recording equipment, separated from an observation/control room by a door and one-way mirrored window. The experimenter described the nature of the investigation and introduced the reader to the text and system. Any questions the reader had were answered before a 3-minute familiarization period commenced, during which the readers were encouraged to browse through the text. After three minutes the readers were asked several questions pertaining to estimated document size and range of contents viewed. They were then given the question set and asked to attempt all questions in the presented order. Readers were encouraged to verbalize their thoughts and a small tie-pin microphone was used to record their comments. Movement through the text was captured by video camera situated non-intrusively directly behind them.

9.2.3 Experimental hypotheses

As the model of likely behaviour deduced from the framework suggests, a single prediction about the most suitable application for these tasks was not possible. Paper would seem to have certain advantages over all electronic versions in some circumstances while it is possible to see advantages for certain electronic versions in others. However, several experimental hypotheses suggested themselves on the basis of the framework.

1. The size of the document and the lack of any specific superstructural model of the information space should convey a general advantage to paper and word processor versions in the first instance. This should manifest itself in problems estimating the document size and greater navigational difficulties in the hypertext conditions.
2. The access mechanisms in the electronic versions should convey advantages to them over paper for certain tasks e.g. those supportable by using search facilities. Therefore the HyperCard and word processor versions should lead to faster completion rates on those tasks than the HyperTIES and paper versions.
3. The verbal protocols should demonstrate the issues of concern to readers at all stages of task performance and will map directly to the elements of the framework.

9.3 Results[3]

9.3.1 Estimating document size

The results of the study generally support part of the first hypothesis: hypertext users had difficulty assessing the document size accurately while readers in the linear conditions were far more accurate. After familiarization with the text, readers were asked to estimate the size of the document in pages or screens. The linear formats contained 13 pages, the HyperCard version contained 53 cards, and the TIES version contained 78 screens. Therefore raw scores were converted to percentages. The responses are presented in Table 9.1 (where a correct response is 100, scores above and below this number reflect over- and under-estimates respectively).

Table 9.1 *Readers' estimates of document size*

Condition	TIES	Paper	HyperCard	W.Processor
Subject				
1	641.03	76.92	150.94	92.31
2	58.97	92.31	56.6	76.92
3	51.28	76.92	465.17	100.0
4	153.84	153.85	75.47	93.21
Mean	226.28	100.0	187.05	90.61
SD	280.41	36.63	189.84	9.75

Readers in the linear format conditions estimated the size of the document reasonably accurately. However, readers who read the hypertexts were less accurate, several of them over-estimated the size by a very high margin. While a one-way ANOVA revealed no significant effect for condition ($F_{[3,12]} = 0.61$,NS) these data are interesting. They suggest that subjective assessment of text size as a function of format is an issue worthy of further investigation and thereby confirm the importance of this issue as indicated in the information model component of the framework.

9.3.2 Navigation

The other part of the first hypothesis was that navigation would pose more difficulties for users of the hypertexts than for the others. As stated in Chapter 3, a general measure of navigation is non-existent but relies on the interpretation and operationalization of the concept by individual researchers. For present purposes it was assessed by examining the proportion of time spent viewing the contents/index (where applicable) by each reader as a percentage of total time. This provided a highly objective behavioural measure rather than any indication of subjective difficulty. These scores are presented in Table 9.2.

This table demonstrates a very large difference between both hypertext formats and the linear formats. A one-way ANOVA revealed a significant effect for condition ($F_{[3,12]} = 9.95$, $p < 0.005$). Even using a more rigorous basis for rejection of the null hypothesis in *post hoc* tests than the 5 per cent level, i.e. the $10/k(k-1)$ level, where k is the number of groups, suggested by Ferguson (1959), which results in a critical rejection level of $p < 0.0083$ in this instance, *post hoc* tests revealed significant differences between paper and

*Table 9.2	Time spent viewing Contents/Index as a percentage of total time*

Condition	TIES	Paper	HyperCard	W.Processor
Subject				
1	53.28	2.72	47.16	6.34
2	25.36	1.49	19.10	13.93
3	49.50	10.24	17.50	12.87
4	30.84	5.36	23.40	7.54
Mean	39.74	4.95	26.79	10.17
SD	13.72	3.88	13.81	3.79

TIES ($t = 4.90$, d.f. $= 6$, $p < 0.003$), between word processor and TIES ($t = 4.16$, d.f. $= 6$, $p < 0.006$) and between HyperCard and paper ($t = 3.06$, d.f. $= 6$, $p < 0.03$). Thus, interacting with a hypertext document may necessitate heavy usage of browsers or indices in order to navigate effectively through the information space.

9.3.3 Searching for precise information

It is expected that when readers seek information for which they can formulate accurate search terms, applications that offer such facilities should lead to faster and/or more accurate task completion rates than those which do not. However, use of the search facilities rests on several factors: the realization of their presence; the willingness to use them; and the ability to use them correctly.

It was expected that the readers in the present sample would be familiar with search facilities and realize that they existed in the relevant applications. The willingness to use them is more difficult to predict and it is possible that users will only employ them when other tactics fail. Using them correctly is a skill and any realistic model of user behaviour must allow for possible errors.

In this study, three of the tasks were supported by the search facilities in the HyperCard and Word conditions. The mean time per reader on these tasks is shown in Table 9.3.

*Table 9.3	Mean times to perform tasks supported by search facilities*

	No Search Facilities		Search Facilities	
Condition	TIES	Paper	HyperCard	W.Processor
Subject				
1	194.33	66.00	112.00	97.00
2	79.67	34.67	44.67	59.33
3	281.67	233.00	39.33	81.33
4	122.67	170.67	36.33	76.33
Mean	169.59	126.08	58.08	78.50
SD	88.43	91.99	36.11	15.57

Analysis of time taken to locate information using the various applications confirms the view that those applications supporting search facilities would be more effective for such tasks. A one-way ANOVA comparing those applications with search facilities and those without showed a significant effect in the hypothesized direction ($F_{[1,15]} = 6.10$, $p < 0.03$) thereby allowing the rejection of null hypothesis two. Interestingly, not all readers used the search facilities for all possible tasks, as suggested above.

9.4 Evidence for the interactive elements from readers' protocols

The data described above demonstrate that the framework can be employed to guide reasonably accurate predictions about reader task performance. However, for present purposes, the main aim of the validity study was to examine the extent to which the framework can be seen as a fair representation of the issues involved in reader–text interaction. To test for this, the verbal protocols of the readers were examined.

Each protocol was transcribed from a videotape according to a pre-defined classification scheme derived by the present author in conjunction with the other members of the research team. This scheme captured the verbal utterances, the time they occurred, the actions performed by the readers and any further behaviours deemed relevant by the evaluator, such as readers having difficulties with an application or making an error in their answer. These were subsequently examined by the author in order to identify verbalizations that mapped onto the framework. Protocol data are rich and complex therefore not easily reduced to a simple presentable form. In the present context isolated sections from a selection of readers in a manner akin to Suchman (1988) seems to be the most appropriate means of highlighting the existence of the elements of the reading process described in the framework. Accordingly several examples from different readers are presented to provide an insight into the reading process from the point of view of the reader.

Example 1 describes a typical section of protocol from one reader. The protocol presentation format involves a time scale in seconds, a transcription of the user's verbal protocol and a description of user action on the system.

This represents a short section (1 minute) of a reader looking for specific information in a HyperCard stack. While on the surface it might seem trivial, it is obvious that much activity is occurring in the reader's mind. Firstly, without hesitation, they have accessed the index. This is logical behaviour as indices provide references to and the locations of material contained in the text. However this implies that not only does the reader have expectations of what they will find in that section of the text but that they have formulated a means of task resolution (or at least a first step towards it) and acted upon it rapidly. Without some form of awareness of the order of the text, the manner in which it might be structured and the access mechanisms available within it,

EXAMPLE 1

TIME	COMMENT	ACTION
00:00	'O.K. I'm going to the Index to see if any of these terms are mentioned . . .'	Selects INDEX Button
00:03	'Don't appear to . . .'	Reading Index
00:25		Selects CONTENTS button
00:35	'I'll have to look in . . .' 'probably section 2, *The Making of Wine* . . .'	Reading Contents
00:58	'I'll go to sweetness because it's the only term in the contents list that really . . . refers to taste . . .'	Selects SWEETNESS

i.e. knowledge of the information space and the manipulation facilities available, such rapid and meaningful processing would not be possible.

It would be virtually impossible to determine the order in which the cognitive processing went for such an interaction. Obviously they knew their task and had a formative map of the document (at this stage the reader had already spent three minutes becoming familiar with the text prior to commencing the trial). These elements must have interacted but whether that interaction was of the form TM→IM or IM→TM is probably not important from a designer's point of view. In reality is was probably a case of cyclical interaction between both. What is certain is that the reader decided on a course of action, manipulated the information space and rejected the first attempt at problem resolution in a matter of seconds.

In fact, in the first few seconds of this example it is possible to see all four elements of the framework in action. Once they decide to go to the index the manipulation element operates (button selection) and the reading processor takes over as they scan down the list of topics to reveal no mention of the target item.

The first interaction of another reader, this time in the word processor condition, is presented below.

Again the interaction of several sub-processes of the reading task can be identified. Upon reading the question, the reader's first decision is to go to a section of the text that might offer guidance pertinent to the task. As before, the expectation that contents sections adequately map the information available in the text, and the fact that they are located in a particular place, should be noted. At this point an interaction between TM and IM has occurred followed by a quick burst of MSF and SRP activity to move up the text on screen and identify the relevant section as the desired one.

Upon reading the contents (SRP activity) this reader identifies two sub-section headings that seem relevant to the representation of the task (TM activity) and decides to go to that section (IM activity) to check if they are

EXAMPLE 2

TIME	COMMENT	ACTION
00:10	'Right, I'm going to the Contents and try and get some idea of where this might be...'	Drags SCROLL bar up
		Reading contents
00:24	'I think it's probably in "The Making of Wine"''	Still reading CONTENTS
00:38	'Oh is it Sweetness and Body?'	Referring to items in CONTENTS
00:42	'I'll just go there and check if they're the two...'	Scrolls down the first then clicks in SCROLL bar until she reaches relevant section. Reads them.
00:58	'So the two things are...' 'Sweetness and Body...'	Finds answer

relevant. At this point their scroll down the page before deciding that the scroll bar would be a better option (MSF activity interacting with TM activity and IM activity – the reader must decide at some point that the information is located suitably 'far away' to justify a faster or more efficient scrolling method). Upon arriving at relevant section (rapid SRP, IM and MSF activity) they cease scrolling and start to read the text (SRP activity) until their opinions are confirmed.

At this stage in both examples only the crudest information model of the text exists. Since this is a unique text type to which the readers would not have been previously exposed, there would be no existing structural models for them to employ and only maps could be formed. With increased exposure this map is elaborated allowing more accurate predictions of what is located where and the type of information available within it. Thus one observes numerous comments to the effect that 'I've seen this before somewhere' or 'I've an idea where this one is...' and so forth. The following example is from a reader using the word processor document who is now on their fifth task.

Here we see the emergence of landmark knowledge of the information space ('I've seen nothing on that'...'I've just passed a section on'...etc.) and knowledge that can allow the reader to make informed judgements about locations that enable them to evaluate (and in this case, reject) hypothetical locations of information ('Grapes on page 1?...No'). Such processing is only feasible where the individual has at least a rudimentary map of where things are in the document, what they are next to, whether or not they have been seen before and what type of information a section may contain. As suggested by the framework, this type of knowledge is likely to be picked up by readers as they become familiar with a document.

Also pertinent here are the limitations of the manipulation facilities for scrolling text in word processors. In going to the index in the first instance this user drags the scroll bar down to the bottom of the window with the result that they overshoot the start of the index section and needs to scroll gently back up (a total activity taking approximately 10 s). This is similar to, but probably more awkward than using indices at the back of books, i.e. overshooting to the back of the book so that you need to page back to the index section is a common experience during reading but should not waste 10 s. Technology should make such a process simpler (or unnecessary), not more difficult. This is an action that is well supported in the 'point and click' facilities common to the hypertext systems.

EXAMPLE 3

TIME	COMMENT	ACTION
06:07	'To the Index then...' 'I haven't seen anything on this before...'	Drags SCROLL bar down
06:09		Scrolls up to top of Index
06:17	'Grapes on page 1?' 'No...'	Reading Index Still reading Index
06:25		Drags SCROLL bar up to Page 2
06:30		Scrolling up to top of section Reads section
06:42		Scrolls down and reads next section
06:56		Scrolls up and rereads section
07:00	'It must be in the body of the report then...'	
07:30		Spends 30 s. scrolling down and reading the following four sub-sections
07:33	'O.K....dessert wines'	Locates answer
07:42		Reads next question
07:45	'I've just passed a section on ageing...'	Scrolls up to that section.

These examples are typical of the protocols elicited in this study. All protocols were examined in this way and it would be overwhelming (not to mention unnecessary for present purposes) to reproduce them all at this level, however a complete protocol for one reader is presented in the appendices to give an indication of the process. The framework proposed here seems to be supported by the evidence from protocols of readers using both paper and hypertext versions of a document in the following ways:

1. There is evidence of the existence of each of the elements, i.e. readers verbalize thoughts which confirm attention changing from attributes of task, information model, manipulation and serial reading of the text.
2. There is evidence of the interaction of the elements in both linear and non-linear fashions (i.e. interactions do not necessarily follow the strictly linear sequence: TM→IM→MSF→SRP but combine in sequences which reflect the reactive nature of the reading process).
3. There appeared no evidence from any of the protocols to suggest other elements need to be incorporated in the framework, all relevant verbalizations and behaviours proved classifiable as belonging to one of these categories (though of course, these categories or elements are rathe general and hide many complex cognitive issues as stated earlier).[4]

It seems as if the framework provides a relatively parsimonious account of the types of utterances solicited from readers performing routine tasks with a text. No attempt has been made here to compare the verbal protocols between conditions to identify any differences that might exist between them. For example, it is possible that readers in the hypertext condition manifested less comments on their information model than readers in the paper condition, or most comments in all conditions were about manipulation rather than scanning issues. While such findings might be of some interest in terms of the psychology of reading they are secondary in importance to the aims of the present work, i.e. the development of a useful descriptive framework of the reading process to aid electronic text designers.

A further reason for not pursuing a more quantitative analysis of the protocol data is that such a procedure is difficult to perform objectively. In the examples described above one can identify the general sequence of activities and their relationship to the framework. However, to force every protocol into an 'all or nothing' categorization whereby each utterance is classified as belonging to one or other element in the framework would hardly be informative and would lead to a mass of numeric data that added little or nothing to the present description of the framework. Many of the interesting utterances concern the rapid interaction of several elements or the continual interchange between two non-adjacent elements. If the quantitative categorization was to take account of all elements, the possible interactions between elements and the time line associated with each utterance and try to relate these to each of the four conditions in the experiment it is likely that the resultant analysis would over-complicate the data to the point of meaninglessness.

In summary, the framework posits the existence of four elements of concern to the reader of any text. The verbal protocols support the existence of these and suggest that the interactions between these elements is of the general form described in Chapter 7. The framework provides an adequate account of the type of processes carried out by readers of both electronic and paper texts.

9.5 *The utility experiment*

9.5.1 Overview

The previous study also showed that the framework can support accurate predictions about reader performance with a text as a function of presentation medium. The present experiment extends this to another text type and employs a further hypertext application (GUIDE) not used in the last experiment.

A hypertext academic journal article is unlikely to be a complete replacement for the paper version. On the basis of the task analysis and literature review it would seem that for straight reading of the text, paper would often be preferred and prove more usable. However, for the other forms of use to which such texts are put hypertext is likely to offer certain benefits. This text type would therefore seem a useful test of the utility of the framework.

9.5.2. Applying the framework to the description of academic article usage

A journal reader approaches a text with a task or set of tasks that they hope to resolve, no matter how ill-specified. According to the framework it is suggested that the readers apply their model of the text structure to the task in order to direct their activities. Thus they decide if they need to look at some part of the article more than the other, where that part is located, where the other relevant articles might be and so forth. If the electronic version maintains the paper structure there should be no differences between the media at this stage, i.e. their well-developed model would be equally relevant to either medium.

Readers then manipulate the text and locate the section(s) relevant to their needs. Traditionally this would have been difficult with electronic text but the availability of hypertext applications eases the manipulation task considerably, particularly where text is broken or 'noded' into selectable chunks. In the case of the article, jumping to various sections and headings should be facilitated on screen, though location of particular blocks of text within a larger body of text is unlikely to be so easy.

Once at the relevant section it is probable that readers adopt one of two reading styles: straight serial reading from the start or quick scanning. In reality, readers probably adopt a mixture of both. Where the reader adopts a serial reading style, paper is likely to be better than hypertext. This seems probable given the weight of evidence showing a performance deficit for proof-reading speed from screens and the difficulties readers have with lengthy electronic texts. However, the differences between the two media are likely to be lessened where the amount of text to be read is small (i.e. a screenful). Where the reader is scanning the material and it is not lengthy, there is likely to be little difference between the media, assuming image quality of the screen is good. If the text is broken into various small sub-sections within a section and the reader has an idea where they want to go, hypertext should convey advantages over paper.

Accordingly a simple descriptive model of user behaviour for a particular task, for example, checking a detail in the method section such as the number of readers employed or the type of equipment used could be derived. In circumstances where the paper article's structure is retained in hypertext no difference would be assumed between the media until the MSF and SRP elements of the framework are invoked. At this point one would expect an advantage to the hypertext version for getting to a headed section but an

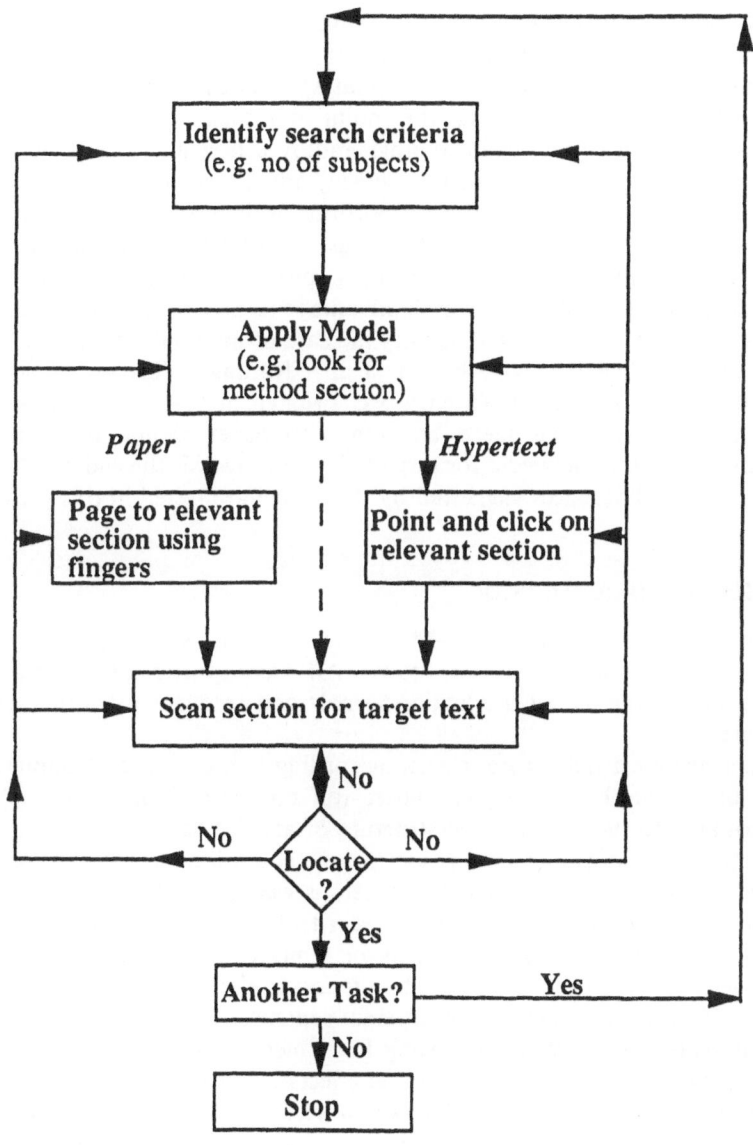

Figure 9.2   A schematic model of readers' behaviour for journal task.

advantage to the paper version for the scanning or serial reading phases of performance. A descriptive model of such a task is represented in Figure 9.2.

According to this model the task basically consists of quick target identification, a rapid application of the IM, and then a sequence of MSF ↔ SRP interactions. The latter interactions dominate the task according to the framework though each element is not used in equal proportion. Given the task involves scanning text in a specified area there is likely to be a bias towards SRP involvement. MSF activity should be rapid while SRP activity could be extended.

From what is known about reading from contemporary screens it is clear that for SRP activity, paper should be faster than hypertext (approximately 20 per cent faster according to reliable estimates). While manipulation can prove problematic for electronic texts, the 'point and click' approach of GUIDE should be familiar to the reader sample employed here and where the targets can be directly addressed from one screen, advantages to hypertext should ensue. However, given the relative proportions of time estimated to be spent in either activity (i.e. the majority of the task is SRP activity, not MSF) this should not be enough to offset the reading speed advantage to paper.

Though there should be an overall advantage to paper it is possible that for targets not requiring large SRP activity the ease of manipulation with GUIDE might prove sufficient to give hypertext an advantage. This would occur in situations where the target sentence was situated at the start of an opened section. Conversely, the speed advantage to paper should maximize the differences between the media for targets situated towards the end of a lengthy section. These differences suggested by the model are tested in this study.

9.5.3 Readers, texts and tasks

Twelve readers (age range 22–35, mean age 27, six male/six female) participated voluntarily in this study. All were professional researchers experienced in the use of academic journals and frequent users of personal computers.

Two academic articles were selected according to the criteria of similarity in terms of length, broad subject matter (on computer human factors) and conformance to the general superstructure of articles described in Chapter 7. Good quality photocopies were made for one condition and GUIDE versions created for the other. The hypertext versions were presented, black on white on an Apple Macintosh II. Screen Recorder[TM] was used to record (non-intrusively) readers' performance with the hypertext.

Readers were required to locate 32 sentences in the academic articles. These were divided into four task blocks of eight sentences each, so that each reader located sentences in both texts using both media (in order to control any possible text biases or presentation order effects). The sentences were presented on stimulus cards which stated in which section of the text (introduction, method, results, discussion) the sentence was to be found. The task was designed to be a simulation of the situation common to readers of these texts which is checking a detail of the paper when they have a fairly reliable notion of where the target sentence is located but are not likely or able to use the search facilities (i.e. they cannot formulate an appropriate search string).

The sentences were selected so that equal numbers came from all four sections, they were of approximately similar length (i.e. less than two printed lines) and did not contain eye-catching words (e.g. all-capitals) or symbols (e.g. numbers). The hypertext versions of the texts were made so that line lengths were comparable to the paper ones thereby ensuring a similar typographical form for both media. As well as being situated in particular sections, the target sentences were further distinguishable in terms of their within-section location. Thus, in all but the method section for which it would be impossible, the sentences in each section were selected so that they were at the start or towards the end of the relevant section. The qualification for being located at the start was determined by the presence of the sentence in the first full screen of text that was presented upon opening a section in the hypertext. This allowed further analysis of readers' performance, i.e. the effect of scanning large amounts of text on either medium.

A two condition (paper × screen) repeated measures design was employed using two texts. All readers performed the task twice on paper (once per text) and twice with hypertext (once per text) with order of texts and presentation medium counterbalanced to avoid any systematic ordering effects. The independent variable was presentation medium and dependent variable was speed of task performance.

Readers performed the experiment in an experimental room at HUSAT. The computer was placed on a desk which was free of other materials allowing them to perform both the paper and the hypertext tasks without changing desks. The investigator sat at the edge of the desk in a position where the screen and the document being read could be seen at all times.

All readers were introduced to the concept of hypertext and the specific workings of the GUIDE package. Most expressed familiarity with the concept but only three participants had actually used GUIDE. All were then encouraged to interact with the application until they were comfortable with it, at which point they performed five trial tasks to consolidate their training. If they still experienced any difficulties further blocks of trial tasks were available. However, no reader asked for, or appeared to require, extra training.

Readers were informed that they would be timed for each individual task. Timing started from the moment the experimenter handed them the stimulus card containing the target sentence until the time when they successfully located it. Successful location was marked by a verbal statement of the fact and pointing to the sentence with finger or cursor, enabling the experimenter to confirm that the target was successfully located. Upon location, the experimenter noted the time lapsed and ensured that the reader closed the paper version or went to the top-level of the GUIDE document before commencing the next task. There were two minute rest periods between each block of trials.

At the end of the experiment readers were asked to describe their general ratings of the hypertext version and its suitability for journal article presentation.

9.5.4 Experimental hypotheses

Given the task and the conditions under which they are presented is was expected that there would be only two levels of difference between the media: the manipulation and the skim reading ones (MSF and SRP elements) as outlined above. Given the variation in locations and the estimated proportion of time spent on each activity, three experimental hypotheses were proposed.

1. There would be a significant difference overall between the two presentation media for the completion rate of tasks with paper proving faster than hypertext.
2. Readers should locate information in lengthy sections of the texts faster with paper than with hypertext.
3. Readers should locate information in short sections of text faster with hypertext than with paper.

9.6 Results

9.6.1 The effects of medium, text and question on performance

A three-way, $2 \times 2 \times 8$ ANOVA (medium by text by question) with repeated measures on all factors was carried out on the data. Although the texts were selected for similarity it was decided to test for text in case it was producing an effect that was not dependent on the variables controlled for in their selection such as idiosyncratic writing style or vocabulary. The output from this analysis package is summarized in Table 9.4.

These results indicate significant effect for medium, question and the interaction between medium and question. There was no significant three-way interaction effect or significant effect for text type as expected.

Table 9.4   ANOVA summary table for utility experiment

Source	d.f.	SSQ	MSQ	f	p
Guide/Paper (A)	1	33189.85	33189.85	14.909	0.0029
Text 1/2 (B)	1	2470.51	2470.51	1.7418	0.2119
AB	1	133.01	133.01	0.2187	0.6524
Question (C)	7	122309.5	17472.79	12.655	0.0001
AC	7	29399.67	4199.952	2.7579	0.0131
BC	7	16546.74	2363.82	1.3393	0.2429
ABC	7	20704.99	2957.856	1.43105	0.2045
Error	352	565404.37	1601.712		
Total	383	790158.64			

Clearly, there was a significant effect for medium with paper proving to be faster than hypertext for this set of tasks. Thus null of the first hypothesis can be rejected. Mean time per task with hypertext was 52 s compared with 33.5 for paper, i.e. paper was approximately 35 per cent faster than hypertext which is just outside the range of speed differences between the media typically reported in the literature.

The significant effect for question was also expected. Searching for targets in text sections of varying length should lead to speed differences with shorter sections affording faster location. This requires no further explanation. Mean times for each location confirmed the direction of the differences, e.g. location times for questions three and four (method section) were 14.65 and 16.46 seconds respectively, while mean location times for questions one and two (introduction section), were 40.77 and 56.60 s respectively.

9.6.2 The effect of target position on performance

According to the framework, readers should have been able to locate sentences that occurred in the short sections of the texts (e.g. in the procedure sub-section of the method section) faster in hypertext than on paper. The opposite should hold for sentences embedded in longer text sections such as the discussion (though where sentences occur at the start of such sections hypertext should regain the advantage). Thus the advantages of either medium should be tested between questions.

Table 9.5 *Mean times (seconds) per question for each medium*

Target	Hypertext	Paper	p
Introduction/early	46	35.5	ns*
Introduction/late	63	50.5	ns
Method 1	15	14	ns
Method 2	17	16	ns
Results/early	75	39	0.05
Results/late	93	42	0.05
Discussion/early	50	46	ns
Discussion/late	57	24	0.05

*ns = not significant

The significant effect for question, and more importantly the significant interaction effect for medium and question give an indication of what happened in the present case. By examining the mean times per task for each medium it is clear that the advantage to paper is most obvious for tasks involving the scanning of large sections of text. No such difference holds in when the target sentence is located in shorter sections. The unweighted marginal means for task by medium are presented in Table 9.5.

P values were obtained by using the marginal means to calculate a value for *t* according to the formula

$$t = \frac{(\text{mean } 1 - \text{mean } 2)}{\sqrt{(\text{mse}/\text{n1} + \text{mse}/\text{n2})}}$$

(Ferguson, 1959, p. 238), where mse = 1522.86 and df = 77(= 1 × 7 × 11).

These data provide a better view of the results than the overall difference between media. Far from being enormously better than hypertext for this type of task it can now be seen that the advantage to paper is maximized mainly for location of material that is situated towards the end of lengthy sections. Although this difference was non-significant for target sentences in the introduction, it was still large and in the hypothesized direction. This supports the second experimental hypothesis for which the null hypothesis can now be rejected. For targets that occurred in the first few paragraphs of a section or in sections that did not contain large expanses of straight text (e.g. the method sections) then readers performed as well with hypertext as with paper. However they did not perform significantly better, as predicted, therefore necessitating the retention of the third null hypothesis.

The significant effect for targets in the early parts of the results section runs counter to the suggestion drawn from the framework. Examining these tasks on the Screen Recorder output confirmed what had been suspected by experimenter observation, i.e. readers regularly missed the target on first exposure to the section and read serially through to the end before returning to the relevant part of the text. In all, eight readers failed initially at least once to locate the target even when it was first on screen. Of these, six failed to do so on more than one occasion and two readers missed the same target on two occasions (i.e. they reread a section more than twice before locating the target). Though there were no equivalent data records of readers in the paper conditions, the experimenter noted only one reader doing this.

9.7 Discussion

The present investigation was intended to simulate the type of task performed by readers when searching for a specific piece of information in a familiar text type. In the academic article situation this would involve searching for a reference, checking a detail in the design, or finding the major results etc. It was hypothesized that this would be the type of reading task for which hypertext would in part, offer suitable, perhaps even advantageous support.

The task consisted primarily of two components of the reading process, manipulation and skimming of texts, which were related to two elements of the framework. Hypertext offered clear speed advantages for getting to a section, it was merely a matter of pointing and going. Though not timed at this level, paper required the use of both hands to flick through the pages and readers often paged past the sections they wanted in their attempt to jump directly to it. This supports the original view of hypertext as a potentially advantageous medium for manipulating large texts.

However, once at a section, readers performed faster with paper, particularly when the target sentence was not immediately obvious. This is explicable, at least in part, in terms of the image quality hypothesis discussed earlier. The statistically significant advantage to paper overall predictably emerged as a result of concatenating all the actions into one performance score: total time. Since the proportional time spent scanning was normally greater than that spent manipulating the text, this weighted the final measure in favour of the task component known to be best supported by paper (in other words, the task lessened the influence of one of hypertext's major advantages).

These results support two out of three of the hypotheses derived from the framework. However, the predicted advantage to hypertext for locating text in short sections of text never materialized. In fact, the general trend of the results has been to distort the predicted differences further in the direction favouring paper than was expected, i.e. the reading speed difference was larger than normally observed. How can this be explained?

Several issues to do with general presentation on screen are worth considering here. Although the hypertext was presented as black on white, the font (New York 12 point) was chosen to retain similarity to the printed paper font rather than optimizing the screen display. On the basis of this study and that of the second experiment on readers' structural models reported in Chapter 7, this was a mistake. Paper fonts are optimized for paper reading, screen fonts should be optimized likewise.

One reader remarked that Geneva would have been a more suitable font and it was clear from talking to participants afterwards that text presentation is generally viewed as poorer on screen than on paper regardless of font. A variety of associated reasons such as angle of viewing, flicker, and other subjective measures which have failed to produce clear explanations of empirical differences in the work of Gould *et al.* (1987a, 1987b) *inter alia*, are still reported by readers as potentially disruptive.

The value of trying to mimic the line length of the paper article on screen was also questioned by two participants. They both felt that for screen reading, a wider display with increased interline spacing would have been more readable and helped them to identify relevant target characteristics more easily. There is some evidence in the literature to support these suggestions (e.g. Kolers *et al.*, 1981; Duchnicky and Kolers, 1983). In terms of the descriptive model derived from the framework there are obviously few, if any, modifications required to explain these findings. It led to the accurate prediction that paper would be better than hypertext both overall and for locating sentences situated in the later parts of lengthy sections. The shortcoming of the other hypothesis seems to result not from shortcomings of the framework or any of its postulated elements but from the task and sensitivity of the measures employed.

The sentence location task was intended to simulate the type of reading scenario where a person searches relatively familiar material to identify a certain detail. Behaviourally at least the experimental task employed here matched this. However, cognitively it is difficult to maintain the comparison.

Readers in the present situation reported trying to locate the required sentence in a simple pattern match fashion, i.e. they focused on a key word or phrase in the target sentence and searched the text closely for anything that matched this without considering the meaning of the material being attended to. In the real reading situation one would expect the readers to be more influenced by the context of the material and seek to relate the content of the currently attended-to paragraph to their information needs.

In this way, real reading would involve the narrowing of the search space according to the context of the target's location, e.g. if a reader wants to find a sentence about difficulties with the experimental procedure they are likely to appreciate the relevance of other words or sentences which mention other problems or procedural issues. There was no evidence of readers in the present study actually trying to relate the content of the target sentence to the experimental text beyond the cues they were given for searching in a specific section. Indeed, the typical reading style manifested by readers was a straight serial read from start to target (or finish) of the prompted section, scanning every intervening word. In reality, one would expect to see readers jumping about within sections, ignoring paragraphs which the first sentence indicated were unlikely to contain the required details. This would have the effect of speeding up this process and thereby lessening the proportion of task time spent at the SRP level of the framework with commensurate benefit for hypertext users.

A further source of potential bias in the study was the assumed lack of training required by readers in using GUIDE. Although all readers expressed confidence after the familiarization period with the use of this application for manipulating the text, and there were no instances of readers being unable to open or close sections of the hypertext during task performance, most reported afterwards that they would require a lot more use of GUIDE before feeling as comfortable with it as they were with the paper texts. Several readers certainly manifested non-optimum use during task performance, e.g. opening irrelevant sub-sections and failing to close them before opening another which resulted in large text sections to scroll through if they went back to the 'start' to reread a section.

The design of the experiment, with its demand that readers complete all tasks failed to allow for any potential speed/accuracy trade off. Since many of the readers overlooked the target on initial exposure and were forced to reread sections, sometimes more than once, their performance (i.e. speed) scores deteriorated rapidly. Had readers been given the option of giving up, the speed differences might not have been so large. Furthermore, for most electronic texts, when faced with a situation where a target is proving evasive, the search facilities are likely to be used. These were not supported in the present task as it was an attempt to simulate the type of interaction where the reader has only an ill-formed idea of the specific details of the search, therefore being more likely to use recognition over recall to aid location. Future work would do well to consider such task effects and select accordingly.

In summary, the data suggest that hypertext can be as good as paper for information location tasks when the reader has (or is given) an accurate information model and is not required to read lengthy sections of text. Where lengthy text sections must be read the advantages of paper in terms of image quality can lead to speed deficits with electronic text on current screens.

9.8 The validity and utility of the framework

The framework was examined here in two studies, involving two different texts and four different electronic applications. Its validity has been tested by parsing verbal protocols of readers into convenient chunks and relating them to the elements in the framework where they have been shown to map adequately and sufficiently. For the purposes of providing a non-complex representation of reader psychology relevant to text use it seems valid.

The utility of the framework was tested by examining the accuracy of predictions derived from the framework. Over both studies, five experimental hypotheses were derived and tested, four of which resulted in the rejection of the null hypothesis. No more need be said of these except that they are strong, i.e. unidirectional, hypotheses. For the hypothesis that was not supported, it is possible that experimental design factors are sufficient to explain the data. There appears to be no need to alter the framework to explain the findings. Thus as a means of predicting the likely performance of readers with electronic texts, the framework has demonstrated utility.

The framework has been used subsequently in a variety of other electronic document design projects in HUSAT, albeit not as explicitly as outlined in this chapter. In the final chapter therefore, attention turns to the envisaged future applications of this framework and the lessons that have been learned in this work for the development of future electronic text systems.

Notes

1. There is a further issue here about the extent to which such a model can be accepted as proof of the process or is just a demonstration of one possible way in which behaviour or output can be caused. This distinction is often labelled 'hard' and 'soft' respectively in the artificial intelligence community and will not be discussed further here.
2. It should be noted that the effective procedure test is not universally accepted by psychologists yet. Furthermore, most psychologists would agree that is not even appropriate for many levels of psychological enquiry. Kline (1988) argues strongly against such testing.
3. Some of the data presented here are used by kind permission of my co-workers at the time Cliff McKnight and John Richardson of the HUSAT Research Institute.
4. Not surprisingly, there were verbalizations that were deemed irrelevant, such as comments to the experimenter regarding the time, the ease or difficulty of particular tasks of quips about the study. However these amounted to a very small proportion of the total data elicited and can safely be considered inconsequential with regard to the framework.

10

Designing usable electronic text: conclusions and prospects

Where is the knowledge we have lost in information?

T. S. Eliot, *The Rock* (1934)

10.1 Introduction

The book set itself the primary aim of examining and subsequently describing the reading process in a manner that would support sensible analysis of the potential rôle of information technology in this process. The framework outlined in Chapter 8 is intended to serve this purpose and in so doing, provide a means of raising awareness of human factors in information usage to designers of electronic text systems, the secondary aim of the book. This, the final chapter will review the work in the light of these aims and suggest areas for future research.

10.2 The framework as a description of the reading process

There are three aspects to this issue of description that are important for the framework: level, scope and context. These are discussed in turn in the present section.

10.2.1 Describing reading at an appropriate level of abstraction

The process of reading has been subjected to continued examination by scientists from a variety of disciplines for over a century now. For all that effort, the process has still to be adequately described by any one discipline. Psychology has led the way in trying to understand the cognitive activities involved, while information science has concentrated on the more pragmatic issues of providing people with access to stored material. Educators, typographers, librarians and sociologists *inter alia* have all applied their discipline's tools and theoretical perspectives, and while collectively, progress can be said to have been made, few researchers of reading would claim to have all the answers.

In the present context, the impact of advanced information technology on the reading process has been identified as an issue worthy of examination. With the impetus provided by electronic text in general, and hypertext in particular, this issue is becoming the focus of much attention and speculation. Current research on reading electronic text was reviewed and found to be both piecemeal and of little direct use to those responsible for designing these tools, primarily as a result of the uni-disciplinary definitions of reading adopted by researchers, the limited tasks examined and the resultant failure of any descriptive framework or model to provide a means of conceptualizing the range of issues involved.

It became clear to us at HUSAT from examining the question of usability with respect to electronic texts that the variance that exists in terms of texts and reading tasks is likely to be of crucial importance in describing the reading process. In order to operationalize these factors in a reader-relevant form, people's perceptions of texts and their characteristic manner of using them has been described. The first stage of which suggests that all texts are describable by readers in terms of three criteria: why they read them, how they read them, and what general type of information they contain.

The utility of these straightforward criteria lies in their ability to distinguish between texts according to usage factors and thereby group material in a form that directly supports examination of the potential role of information technology in their use. This sets the classification criteria apart from any other text typology, the majority of which have attempted to classify material in terms of linguistic structures from which mappings to electronic text design are difficult to make. Furthermore, they reflect the view that texts exist in the reader's world and possess affordances to which the experienced reader becomes attuned.

When a text has been conceptualized in these terms one has a basic orientation from which to proceed in further describing the reading process. Thus, given any text, the three criteria can be used to elicit detailed information from readers on the type of tasks it is used for, their manner of interacting with it and the context of typical use. This can be done directly by a researcher or designer based on common-sense reasoning or, as shown in Chapter 6, more objectively through structured interviews and simulated task performance with readers.

In this way, the reading process is initially conceptualized in terms of the text and task involved, hence the initial element of the descriptive framework: the task model. This immediately distinguishes it radically from cognitive psychological analyses of reading which in many ways can be seen as text and task independent. It also distinguishes the description from the type of conceptualization offered in information science which is concerned with the range of texts as categorizable entities but offers little insight into how individual readers actually construe or use them once they have been located.

The reliance on task analysis as a function of text classification promotes a level of description that can be seen as predominantly psychological in its concepts, yet is atypical (in its breadth) of traditional psychological descriptions of reading. In relative terms it is a higher level of description than that provided by cognitive psychology but a lower level of description than that typically provided by information science. This is obvious from two other components of the descriptive framework: the information model and the manipulation facilities and skills element.

The concept of an information model is well-established in the psychological and linguistic literature but tends to be used only as a theoretical construct in discussions on reading comprehension (e.g. van Dijk, 1980 or Garnham, 1987). The link between this work and the more traditional research on reading is only infrequently made. However, the interviews with readers frequently carried out in the course of this work confirm that the concept is inextricably interwoven with text usage, providing a reader with the means of grasping the organization of material as well as supporting accurate prediction of the location of material in a text. The experimental work in Chapter 7 on academic journal articles, and extracts from the verbal protocols of readers interacting with a unique text in Chapter 9 lend support to these views.

The manipulation element is perhaps the least likely component of the descriptive framework. Few people, when discussing reading, ever consider the issue of document manipulation to be of central (if any) importance. However, from the literature on reading from screen reviewed in Chapter 3, it seems that manipulation issues are crucial to the analysis of electronic text. Much reading involves manipulation by virtue of the presentation media humans have developed. From pamphlets to ledgers, letters to novels and manuals to encyclopaedias, document usage invariably requires the reader to open and turn pages, keep fingers in the text portion of interest while opening other sections and so forth. In fact, it is such an inextricable part of the process that without the ability to manipulate material easily, much reading would not be possible (or at best, would prove difficult) with current print media. The framework recognizes the importance of these activities by including a manipulation element in its structure.

The lowest level of the framework represents behaviour more usually equated with the activity of reading. The serial reading element is the component that covers the process of extracting the message from the text, i.e. it refers to the contact or interaction between eye and print, so to speak. When an individual actually examines the text at the word or sentence level, the type of activities common to traditional psychological models of reading such as eye movement, word recognition, lexical processing and so forth are presumed to occur. From the point of view of the framework, these issues are pertinent, but only after, or in conjunction with, the range of behaviours and processes described in the other elements.

The framework therefore describes reading as a task-driven activity involving the setting of goals, the evolution and application of an information model, the manipulation of a document and the visual processing of text images. This is in contrast to the definition of reading as the visual and cognitive processing of textual images typical of psychological textbooks or as the acquisition and usage of documentation, to put it in information science terms. It does not suggest that these are the only valid issues that can be described as reading, nor does it imply that any one of these is more or less important in the whole process. Furthermore, it does not suggest that traditional research paradigms on reading are wrong. Its intention is purely to provide a level of discourse appropriate to the examination of reading in the context of information technology.

10.2.2 The scope of the framework

Each of the elements in the framework raises an issue or set of issues to be dealt with in the design of electronic text. Thus the reading task must first be understood in the terms of the text type and its context of use. The information model element focuses attention on the reader's representation of the document's structure. The manipulation element highlights the importance of such facilities while the serial reading element raises the issues associated with visual ergonomics. Issues that do not map onto one or other of these elements are, according to this framework, of secondary importance to the design of electronic texts.

This latter point is worth elaborating. No scope for the *explicit* analysis of the reading outcome is provided by this framework. So, for example, the concept of comprehension, amongst others, is not represented in the framework; yet comprehension is, for many theorists, a crucial component of reading. This is not a return to the theoretical debate on the appropriateness or otherwise of comprehension in the discussion of reading with a statement of the present author's recommendation to exclude it. Rather it is a reflection of the goal of the descriptive framework: to support the accurate examination of human factors issues in electronic text design.

If technology is designed appropriately, users will be able to gain access to well presented information in an efficient and easy manner. At this point, it is not clear what more can be done with the technology to ensure readers actually make sensible usage of this material, i.e. achieve their goal, find their reference, comprehend the text and so forth. This choice of outcome exclusion emphasizes the general paucity of applicable knowledge available from work on comprehension and similar concepts even for designers of paper texts, which implies that attempting to design electronic text that ensures greater comprehension of material, for example, is not an immediately attainable goal (despite the claims of the technocrats). Obviously, as cognitive science progresses, such goals might become more feasible. They are certainly desirable. However, the present author's view is that currently, they are not practicable, in the sense that a design process cannot be specified sufficiently to ensure their attainment.

Furthermore, it is highly unlikely that one level of description can hope to encompass all possible issues. The nature of scientific investigation is that it divides the world up into levels of analysis. Unified theories are rare (even within religions) and classical science divides itself into the disciplines of physics, chemistry and biology, none of which attempts sweeping explanations outside its accepted boundaries. It should not be expected therefore that given current knowledge, one could devise a single level description of such a complex human activity as reading.

Until we have sufficient knowledge about the relationship between information presentation and subsequent learning or comprehension, then the efforts of electronic text designers should be concentrated on providing the tools to access and manipulate relevant material in a suitably usable manner. This is not defeatist or pessimistic however, and is certainly not equivalent to Clark's (1983) position that media cannot be considered a determining factor in learning outcome. The attainment of comprehension or other outcomes are likely to be contingent upon such successful and easy access provided by well-designed systems, i.e. usable systems are likely to result in greater (or at least faster) comprehension than badly designed ones. In this sense the consideration of such issues is not dismissed but is placed in perspective. Media can have an effect on learning in this way. The reading process as described in this framework is surely a prerequisite to any desirable outcomes such as comprehension. The present framework's exclusion of such concepts from its description of immediately relevant issues is not a dismissal of them but a recognition of their complexity. Put simply, one would not expect a specification of a product to state that it must be built so as to ensure greater comprehension. Though this might be desirable or even required, the specification would state it in more concrete terms, e.g. the system must be faster, more accurate, etc., variables which are presumed to reflect, lead to, or correlate with comprehension.

Interestingly, not all cognitive scientists even consider comprehension to be an issue worth addressing. For example, van Dijk and Kintsch (1983, p. 260) state:

> there is no unitary process 'comprehension' that could be measured once and for all if we could but find the right test. Comprehension is a common-sense term which dissolves upon closer analysis into many different sub-processes. Thus we need to construct separate measurement instruments for macroprocesses, knowledge integration, coherence, parsing.... Comprehension is just a convenient term for the aggregation of these processes: it is not to be reified, not to be tested for.

Ergonomically designed electronic media may well impact differentially on some of these processes and thereby affect the learning process. The crucial point for educators is to know which ones and how.

10.2.3 The framework as context provider for research findings

The framework can also be seen as an aid to understanding the human factors literature on reading. As described in Chapter 3, this literature is replete with empirical studies on issues such as the effect of image polarity, scrolling versus paging, large versus small screens and so forth. Interpretation of the various findings can prove problematic and there are contradictions in findings that cannot be resolved without reference to contextual factors.

The framework offers such a context within which to assess the findings of any one experiment. Thus, when one is presented with the question of optimum screen size and notes the Elkerton and Williges (1984) finding that there is no significant difference between screen sizes of five lines and anything larger and contrasts this with the conclusion reached by Dillon *et al.* (1990a) that screens of 60 lines result in significantly different manipulations than, and are preferred over, 20-line screens, the framework supports the interpretation of each of these findings in an appropriate context by suggesting how variables relating to tasks, texts, manipulation etc. must be considered. Likewise, when Gould *et al.* (1987b) claim that reading from screen can be as fast as reading from paper, the framework supports an interpretation of this statement that allows an informed (i.e. non-literal) acceptance. In this case the framework highlights the fact that electronic text can be as fast for proof-reading short texts on an ergonomically optimized screen. However, this does not mean that no speed deficits occur for other tasks or texts even with such optimized screens. The simple heuristic therefore is: for any statement about the advantages or disadvantages of electronic text, consider its reflection of each of the four elements in the framework. If it misses one (i.e. fails to include assessment of each element) then it is incomplete.

The issue of statement completion is worth elaborating. To make a complete statement about electronic text, reference must be made to the task, the text, the reader's model, the type of manipulation facilities available and the visual ergonomics. For example, the statement:

'paper is better for proof-reading tasks than electronic text'

is more complete than:

'paper is better than hypertext'

but less complete than the statement:

'for proof-reading a familiar text form, paper is better than electronic text'

Though all of these are less complete than the statement:

'for proof-reading a familiar text form, on a typical screen with scrolling facilities, paper is better than electronic text'

In each of these cases the references to particular components of the framework are easily seen, the differences between them lying in the number of elements explicitly articulated. However, despite a statement's completion, its truth content is another factor. A statement may be complete in the sense implied here, but be wrong. However, this is a separate issue. A complete statement is open to evaluation, either in terms of current knowledge or empirical investigation. An incomplete one cannot be so easily tested. For example, the final statement above, is easier to comment on appropriately or to test empirically as valid than the first statement. It befalls researchers and designers alike therefore, when making claims about electronic text, to do so in as complete a fashion as possible. Similarly, incomplete statements (e.g. 'hypertext is superior to paper' or vice versa) must be dismissed as nonsense.

10.3 The framework as a guide to design

10.3.1 Why a qualitative framework?

At the outset it was stated that a secondary aim of the book was to ensure that any resulting description could be packaged and presented in a form suitable for use at the earliest stages of design. It is intended that the framework as outlined in Chapter 8 also satisfies this criterion.

The delivery format of a qualitative framework rather than any alternative such as a set of guidelines or a quantitative model was adopted for a variety of reasons. Predominantly, the emerging perspective on reading was not amenable to reliable quantification. There are few aspects of reading and information design that are amenable to such analysis and the reception of other quantitative models of HCI (e.g. the GOMS model of text editing) by the design community at large hardly inspires confidence in their applicability.[1]

Guidelines were not adopted as there are several problems with them that are well documented, not least their inherent contradictions and over-generalizations. As Hammond *et al.* (1987, p. 41) put it:

> If behaviour results from an interplay of factors, so will the ease of use of an interface. These interdependencies are hard, or even impossible, to capture in simple statements. A guideline which is true in one context may well be misleading in another...the more complex the interface, the less plausible it is that guidelines will help.

The qualitative framework is seen as a suitable alternative to both the standard models or guidelines approaches. It represents a stylistically simple way of presenting a set of complex ideas and supports 'unpacking' of the components to facilitate more detailed analysis. By presenting reading as the interaction of a small number of elements it focuses attention on the range of issues to be considered and their possible interrelationships.

The term 'unpacking' is meant to imply that other forms of advice could be derived from a framework such as this. Guidelines for example could be 'unpacked' from particular components, e.g. 'when transferring paper to hypertext, retain the useful structural components of the original' could be a guideline derived from the IM element, or 'for presenting text on screen ensure image quality is high' could be similarly derived from the SRP element.

Alternatively, existing guidelines could be interpreted in the light of the framework to ensure contextual issues are addressed (thereby lessening one of the major shortcomings of guidelines – their overgeneralized form). For example, the guideline: 'when displaying text that will not fit on a single screen, then use paging rather than scrolling' (Rivlin *et al.*, 1990) if applied rigidly, would lead to some very unusable designs. But if interpreted in the context suggested by the framework, i.e. for certain users, doing particular tasks with specific texts, it is unlikely to be followed slavishly (and, ultimately, inappropriately) by a well meaning designer.[2]

10.3.2 Potential uses of the framework in design

As well as being the most suitable presentation format, the framework is intended to support several uses. First, a designer could use it simply as a checklist to ensure that all important components of the text under design are considered. This guards against the reliance on research findings at one level to ensure good design (e.g. just following the advice on visual ergonomics which concludes that certain fonts, polarity and resolution variables can overcome the reading speed deficit). While that advice might be pertinent and applicable, the framework would suggest that it is but one part of the design problem.

Second, it could be used to guide design by allowing a designer to conceptualize the issues to be dealt with in advance of any specification or prototype. In this sense its advocated use is as an advance organizer (Ausubel, 1968), enabling the designer to organize his thoughts on the problem and highlight attributes of the specification that need to be considered. As discussed in Chapter 8, such an application could lead to significantly more appropriate first specifications or prototypes, lessening the number of iterations required and thereby reducing the time and costs involved in design.

Third, the framework supports the derivation of predictions about readers' performance with a document. The uses made of the framework in the previous chapter highlights its potential value as a predictive tool for a human factors practitioner, adequately familiar with the research in this area, to predict the type of problems a reader will face using an electronic document. It is the author's view that all of the predictions made were easily derived from the framework through the analysis of the various elements and their manifestation or support in the relevant designs, and that few practitioners would face difficulties deriving similarly accurate predictions in other text/task environments.

Finally, the framework has potential evaluative applications. It could be used to guide the expert evaluation of a system under development (i.e. a usability assessment) and support troubleshooting for weaknesses in design. This proposed use is not unlike the first use outlined above except it occurs at a different stage in the design process and is intended to support reasoned examination of the quality of an instantiated design. In this role, one could imagine a designer using the framework to check how the system rated on variables such as image quality, the information model it presents, the type of tasks it will support or manipulations it enables.

10.3.3 ADONIS: a reprise

In Chapter 4, the ADONIS document supply system was described and used to highlight many of the shortcomings the author felt to be present in the literature professing to advise designers, i.e. the system reflected piecemeal application of some guidelines while totally missing some important aspects of the reading scenario that tend not to be covered in the literature. It is reasonable to ask therefore if the proposed framework would have been of any more use to the ADONIS designers.

It goes without saying that this question cannot be answered entirely satisfactorily, without giving the framework to the designers and letting them specify an alternative system while controlling for any further knowledge they might have acquired in the interim. However, one can openly speculate on how ADONIS might have been different if the design team had considered the framework.

Given that the major problems with ADONIS lay with its poor manipulation facilities, it is tempting to assume that even a cursory examination of the issues presented in the framework would have led to more consideration of these in the design team. Considering that the someone involved in the design was obviously concerned enough about image quality to provide a high resolution, black on white screen (for the most part) the framework would have made it clear that this was only one level of issue out of four and therefore, some attention should be paid to the other levels. Had this been done, it is unlikely that a design which restricted manipulation so much, or made searching so awkward would have emerged.

Obviously this is speculation and is not worth pursuing further, but it provides an image of how this framework might be employed. The earliest stages of creative thought are not well understood. In the time it takes to produce an initial idea or respond to a request for a specification the designer must apply a body of knowledge that is a mixture of intuition, experience, stored facts and opinions (Dillon and Sweeney, 1988). If the framework can lead to appropriate inputs to even one of these knowledge sources than it is likely to have had some benefit, and a reader utilizing an electronic text at some future time will reap the rewards of a usable system.

10.4 Further research

Several aspects of reading electronic documents have been highlighted as worthy of further investigation. In particular, not enough is known about the characteristic manner of reading involved for particular texts or text/task combinations. The criteria outlined in the repertory grid study and subsequently applied to the analysis of journals and manuals could be usefully employed to this end. Such a classification of a wide variety of texts would be useful and interesting for those concerned with electronic text design.

A specific topic of relevance is the existence of information models for the type of texts likely to find their way into the electronic medium. The experimental work reported here has concentrated mostly, but not exclusively, on academic literature (and a specific subset thereof), yet it is argued that the concept of structural models is relevant for a multitude of texts. This would then be a natural area for further work and one that cannot be avoided if hypertext is to become anything more than a research and academic curiosity which, one could argue, is all it really is at this time.

In terms of basic human factors work there is still a lot to be learned about screen ergonomics for electronic text, particularly in some of the areas outlined in the literature review such as screen size, manipulation facilities, search facilities and icon design. These are all current research themes in the human factors discipline but some specific work aimed at the electronic text domain would be useful as it is not always clear how findings from one area of application (e.g. visual search) transfer to another.

In contrast to these areas which are ongoing concerns of the hypertext and human factors research communities, a descriptive framework needs to be tested with other designers building a variety of text systems. This would serve to identify not only its particular value but the utility of qualitative frameworks in general for hypertext design. Unlike the other investigations of various reader–text issues outlined above, this is a task that could not be dealt with by simple experimental means or short studies. The type of knowledge that is required would ideally require the investment of large amounts of time and resources, of the type more normally given over to government funded projects under initiatives such as ESPRIT or RACE.[3] In the absence of such resources the framework could be tested through a number of small scale projects with different designers.

10.5 Specifying the design process for hypertexts

At the end of a book such as this, it is justifiable to ask if the work could be summarized into some applicable advice for designers. While it is the intention of the author that the framework should fulfil this requirement, a more explicit statement of how electronic text should be designed in accordance with user-centred principles is probably required. This final section provides a design sequence that involves the framework's components and offers, on the basis of the author's experience, a good chance of successful goal attainment.

Designing a usable hypertext database therefore involves the following stages:

- stakeholder identification and user analysis;
- task analysis of the text(s) or document(s) involved according to three (at least but not exclusive) criteria – **how** it is used, **why** it is used and **what** readers perceive the information to be;
- investigations of the extent to which the document structure is fixed by existing readers' models;

- determining the electronic structure by considering the readers' existing models, potential models and the tasks being performed;
- considering the manipulation facilities required for basic use and ensuring that readers can at least perform these activities simply with the mechanisms provided;
- attempting to add value to the system by offering facilties to perform desirable or advantageous activities that are impossible, difficult or time-consuming with paper;
- ensuring image quality is high; and
- testing the system on users performing real tasks and redesigning accordingly.

The first two steps are important and will provide information of direct relevance to the next three steps. The last step is probably the most important although it is often seen as a luxury that cannot be afforded. Failure to test the design is bound to lead to problems as no theoretical models or formal guidelines exist than can even approach the quality of information obtained from observing real users interacting with a system. It is worth repeating that

'no theoretical models or formal guidelines exist that can even approach the quality of information obtained from observing real users interacting with a system.'

This applies to the descriptive framework proposed in this book as much as to any existing model in the HCI literature. Formal modelling and guidelines have a place at certain steps e.g. tests of manipulation facilities could be performed using GOMS type analyses and guidelines on various features could be employed once the relevant context of use has been established but these are on their own not enough to ensure an acceptable design. These steps represent a complete process and while they will not guarantee success, they offer better prospects of achieving it than many others.

10.6 General conclusion

In 1908, Edmund Huey wrote that to understand reading would be the acme of the psychologist's achievements. That statement is perhaps seen to be more accurate with each successive generation of research on the subject. The subtlety and complexity of the reading process makes it a taxing problem for anyone intent on examining it.

The present work has carved out but part of the reading process as its subject matter. In so doing it has drawn on the ideas and themes of several disciplines concerned with reading. While it might have appeared critical, particularly of the work in cognitive psychology, it is within the psychological perspective that the present work most appropriately lies. It cannot claim to have explained the process satisfactorily, or to have solved any of the thorny issues of what humans do when they read texts. However it has led to a perspective; one that aims at improving the quality of the reading process and ensuring that technology does not make us read despite itself, but actively supports us in this quintessential human activity. The question is not 'should documents be paper or electronic?' but 'how can any presentation medium best satisfy an information need?' There is no simple answer but human factors can help us better understand the question in order to seek the best answer.

10.7 The prospects for electronic text

Groucho Marx once said of television that we call it a medium because nothing is ever well done. In many ways it seems as if the rôle of human factors studies of electronic text has been to adopt a similarly Marxist perspective in highlighting problems with the technology, to show that paper is inherently better or to criticize designers and advocates of the electronic medium for overlooking the human issues. It is hoped that this book has not presented a singularly negative view of electronic text but a realistic one, tempered with the optimism that comes from the author's belief that good design is both possible and beneficial. This section, the final one of the book, examines the prospects for electronic text in the light of the work reported.

Almost ten years have passed since Jonassen (1982) uttered the memorable (and now punishable by quotation!) phrase: 'in a decade or so, the book as we know it will be as obsolete as is movable type today' (p. 379).

Whatever the facts about movable type in 1982, the book as we know it is certainly far from obsolete in the early 1990s. But Jonassen is not alone; the advent of hypertext and desk-top computing means that his point of view is considered visionary in some quarters and that the truth of his claim lies not in its time-scale but its implications.

The implications of widespread electronic text 'any year now' are important. As this work has attempted to highlight, documentation is everywhere: at home in the form of anything from instructions for operating microwave ovens to the novels that induce sleep; at work in the form of texts ranging from reports on latest developments in company sales to the memos that descend from above; and in the world at large in the form of newspapers, advertising boards, shop catalogues and so on. Avoiding documentation in contemporary industrial societies would be a feat of Herculean proportions. Modifying documentation therefore, by presentation in electronic rather than paper forms, will undeniably have an impact on our lives.

In this light the zeal of advocates is understandable, it is just that when humans enter into the equation, accurately predicting these impacts becomes difficult, if not impossible. Paper is familiar, is well liked, easy to use (most of the time), affords a representation of its structure that is quickly acquired by readers and leads to the emergence of conventional forms, is portable, supports excellent image quality and is cheap since publishers have long since recovered their capital outlay on production equipment. Obviously examples could be found of paper documents which flout such conventional benefits but they hold true for the majority of paper texts, while currently the reverse tends to hold for electronic ones.

The progress of electronic text will be neither explosive nor all-embracing. It will only progress by gaining footholds in small task and text domains and by being found usable there (and possibly, at first, only by a few enthusiasts in these domains). As technology develops, screens improve, portability increases and resistance is lowered, the scope for electronic text will broaden, but there is little reason to believe paper will become obsolete in the near future (if ever).

The process will be accelerated by good design, of the kind advocated here, but conversely, it will be hampered by weak design, i.e. that which fails to consider all elements of the framework. It is unlikely that there is anything inherently constraining in the concept of electronic text that cannot be solved by technological improvements and increased knowledge of human information usage. However, the process of reading is not simple and texts are used in multiple ways for myriad tasks by millions of people. Girill and Luk (1983) produced evidence to the effect that for every one line of text read on screen, over one hundred were printed out but this had dropped to a 1:17 ratio three years later (Girill *et al.*, 1986). Even as the ratio continues to drop the amount of paper printed out increases and it should be seen as cautionary that the most successful applications of information technology to date have been the word processor and the photocopier.

Nowadays, the computing world is full of claims and predictions for the latest gadget, the current one being the Apple Newton and equivalent personal technologies. One interesting prediction at the time of writing is that in a few short years we shall all be reading full-length texts with these wonders of mini-engineering and it strikes me on reading such foolish claims, that some technocrats just never learn. Perhaps the only reasonable prediction that can be made is that we shall witness the emergence of dual-form documents: electronic versions for some tasks, paper versions for others. The strengths of the computer will enable cheap storage and rapid access while the intimacy and familiarity of paper will be retained for detailed studying and examination of material.

A text without a reader is worthless. Similarly, a technology without a user is pointless. The human is the key; only by relating technologies to the needs and capabilities of the user can worthwhile systems be developed. The work in this book is a step in that direction for electronic texts, but there is a long journey ahead.

Notes

1. The obvious exception here would be the visual ergonomics issues for which standards on resolution, luminance etc. can be stated quantitatively. Particular aspects of manipulation might also be quantifiable (Card *et al.*, 1978). However, these are very specific instances of HCI that are not singularly concerned with reading.

2. The Rivlin *et al.* guidelines are a prime example of the problems inherent in such advisory formats. While they provide generally useful information to designers, the published set contains at least two erroneous suggestions and several, like the cited example, which sound authoritative but generally fail to allow for important contextual variables which negate their recommendation. For a further review of these guidelines see Dillon (1990).

3. The ESPRIT funded project HUFIT (Human Factors in Information Technology) was one such project (Galer and Taylor, 1989). It ran for five years, involved several major IT suppliers, cost more than £1 million and was intended to produce human factors tools for designers. Interestingly, the actual goal of the project was to design and deliver such tools. The uptake of them amongst IT companies outside of the consortium was considered beyond the scope of such a project, which gives an indication of the amount of work required to input such tools into real-world design teams.

References

Anastasi, A., 1990, *Psychological Testing*, 6th Edn., New York: Macmillan.

Anderson, J., 1983, *The Architecture of Cognition*, Cambridge MA: Harvard University Press.

Annett, J. and Duncan, 1967, Task analysis and training design, *Occupational Psychology*, **41**, 211-21.

Archer, L., 1965, *Systematic Method for Designers*, London: The Design Council.

Asimow, M., 1962, *Introduction to Design*, Englewood Cliffs: Prentice-Hall.

Askwall, S., 1985, Computer supported reading vs reading text on paper: a comparison of two reading situations, *International Journal of Man–Machine Studies*, **22**, 425-39.

Ausubel, D. P., 1968, *Educational Psychology: a Cognitive View*, New York: Holt, Rinehart and Winston.

Bailey, W. A., Knox, S. T. and Lynch, E. F., 1988, Effects of interface design upon user productivity, *Proceedings of CHI, 1988*, New York: ACM 207-12.

Bannister, D. and Fransella, F., 1971, *Inquiring Man*, Harmondsworth: Penguin.

Bannister, D. and Mair, D., 1968, *The Evaluation of Personal Constructs*, London: Academic Press.

Barber, P., 1988, In favour of theory, *Ergonomics*, **31**(6), 871-2.

Bartlett, F. C., 1932, *Remembering*, Cambridge: Cambridge University Press.

Bauer, D. and Cavonius, C. R., 1980, Improving the legibility of visual display units through contrast reversal, in Grandjean, E. and Vigliani, E. (Eds) *Ergonomic Aspects of Visual Display Terminals*, London: Taylor and Francis.

Bauer, D., Bonacker, M. and Cavonius, C. R., 1983, Frame repetition rate for flicker-free viewing of bright VDU screens, *Displays*, January, 31-3.

Beeman, W., Anderson, K., Bader, G., Larkin, J., McClard, A., McQuillan, M. and Shields, M., 1987, Hypertext and pluralism: from lineal to non-lineal thinking, in *Proceedings of Hypertext '87*, University of North Carolina, Chapel Hill, 67-88.

Beldie, I. P., Pastoor, S. and Schwartz, E., 1983, Fixed versus variable letter width for televised text, *Human Factors*, **25**(3), 273-7.

Binder, A., 1964, Statistical theory, in Farnsworth, P., McNemar, O. and McNemar, Q. (Eds) *Annual Review of Psychology 1964*, **15**, 277-310.

Brand, J. and Judd, K., 1993, Angle of hard copy and text editing performance, *Human Factors*, **35**(1), 57-69.

Brems, D. and Whitten, W., 1987, Learning and preference for icon-based interfaces, in *Proceedings of the 31st Annual Meeting of the Human Factors Society*, Santa Monica CA: Human Factors Society, 125-9.

Brewer, W., 1987, Schemas versus mental models in human memory, in Morris, I. P. (Ed.) *Modelling Cognition*, London: John Wiley and Sons, 187-97.

Brown, P., 1988, Hypertext: the way forward, in Van Vliet, J. C. (Ed.) *Document Manipulation and Typography*, Cambridge: Cambridge University Press, 183-91.

Buckley, P., 1989, Expressing research findings to have a practical influence on design, in Long, J. and Whitefield, A. (Eds) *Cognitive Ergonomics and Human Computer Interaction*, Cambridge: Cambridge University Press, 166–90.

Buckingham, B., 1931, New data on the typography of textbooks, *Yearbook of the National Society for the Study of Education*, 30, 93–125.

Buie, S. and Hassell, B., 1982, *An Introduction to Wines*, Electronic document delivered with HyperTIES Software.

Bush, V., 1945, As we may think, *Atlantic Monthly*, 176(1), 101–108.

Cakir, A., Hart, D.J. and Stewart, T.F.M., 1980, *Visual Display Terminals*, Chichester: John Wiley and Sons.

Card, S., English, W. and Burr, B., 1978, Evaluation of mouse, rate-controlled isometric joystick, step keys and text keys for text selection on a CRT, *Ergonomics*, 21, 601–13.

Card, S.K., Moran, T.P. and Newell, A., 1983, *The Psychology of Human–Computer Interaction*, Hillsdale NJ: Lawrence Erlbaum Associates.

Carroll, J., 1984, Minimalist design for active users, in Shackel, B. (Ed.) *INTERACT '84*, Amsterdam: North-Holland, 39–44.

Carroll, J., 1990, Infinite detail and emulation in an ontologically minimized HCI, in *CHI '90*, New York: Association of Computing Machinery, 321–7.

Carroll, J. and Campbell, R., 1986, Softening up hard science: a reply to Newell and Card, *Human-Computer Interaction*, 2(3), 227–49.

Catterall, B., Allison, G. and Maguire, M., 1989, HUFIT: Specification and Design Tools, in Megaw, E.D. (Ed.) *Contemporary Ergonomics '89*, London: Taylor and Francis, 97–102.

Chalmers, A., 1976, *What is this thing called science?* Milton Keynes: Open University Press.

Chapanis, A., 1988, Some generalizations about generalizations, *Human Factors*, 30(3), 253–68.

Chapman, L.J. and Hoffman, M., 1977, *Developing Fluent Reading*, Milton Keynes: Open University Press.

Charney, D., 1987, Comprehending non-linear text: the role of discourse cues, in *Proceedings of Hypertext '87*, University of North Carolina, Chapel Hill, 109–20.

Clark, D., 1983, Reconsidering Research on learning from Media, *Review of Educational Research*, 53(4), 445–59.

Cohen, G., 1988, *Memory in the Real World*, London: Lawrence Erlbaum Associates.

Coltheart, V. and Evans, J., 1982, An investigation of semantic memory in individuals, *Memory and Cognition*, 9(5), 524–32.

Conklin, J., 1987, Hypertext: an introduction and survey, *Computer*, September, 17–41.

Creed, A., Dennis, I. and Newstead, S., 1987, Proof-reading on VDUs, *Behaviour and Information Technology*, 6(1), 3–13.

Cross, N., 1985, Styles of learning, designing and computing, *Design Studies*, 3, 157–62.

Crowder, R., 1982, *The Psychology of Reading: An Introduction*, New York: Oxford University Press.

Cuff, R., 1980, On casual users, *International Journal of Man–Machine Studies*, 12, 163–87.

Curtis, B., 1990, Empirical studies of the software design process, in Diaper, D., Gilmore, D., Cockton, G. and Shackel, B. (Eds) *INTERACT '90*, North Holland: Amsterdam, xv–xxi.

Cushman, W.H., 1986, Reading from microfiche, VDT and the printed page: subjective fatigue and performance, *Human Factors*, 28(1), 63–73.

Darke, J., 1979, The primary generator and the design process, *Design Studies*, 3, 157–62.

de Beaugrande, R., 1980, *Text, Discourse and Process*, Norwood NJ: Ablex.

de Beaugrande, R., 1981, Design criteria for process models of reading, *Reading Research Quarterly*, **16**(2), 261–315.

Diaper, D., 1990, Simulation: stepping stones between requirements and design, in Life, M., Narborough-Hall, C. and Hamilton, W. (Eds) *Simulation and the User Interface*, London: Taylor and Francis, 59–72.

Dillon, A., 1987, Knowledge acquisition and conceptual models: a cognitive analysis of the interface, in Diaper, D. and Winder, R. (Eds) *People and Computers III*, Cambridge: Cambridge University Press, 371–9.

Dillon, A., 1988a, 'The ADONIS Document Delivery Workstation: a user interface evaluation', Project Quartet Deliverable, HUSAT Research Institute, Loughborough University of Technology, Leics. England.

Dillon, A., 1988b, 'The role of usability labs in systems design, in Megaw, T. (Ed.) *Contemporary Ergonomics, 1988*, London: Taylor and Francis.

Dillon, A., 1990, A review of Rivlin *et al.* (1990): Guidelines for screen design, *Hypermedia*, **2**(2), 171–3.

Dillon, A., Richardson, J. and McKnight, C., 1989, The human factors of journal usage and the design of electronic text. *Interacting with Computers*, 1(2), 183–189.

Dillon, A., 1991a, Requirements analysis for hypertext apllications: the why, what and how approach. *Applied Ergonomics*, 22(4), 458–462.

Dillon, A., 1991b, Readers' models of text structures: the case of academic articles, *International Journal of Man-Machine Studies*, 35, 913–925.

Dillon, A., 1992, Reading from paper versus screens: a critical review of the empirical literature, *Ergonomics: 3rd Special Issue on Cognitive Ergonomics*, 35(10), 1297–1326.

Dillon, A. and McKnight, C., 1990, Towards a classification of text types: a repertory grid approach, *International Journal of Man-Machine Studies*, 33, 623–636.

Dillon, A., Richardson, J. and McKnight, C., 1990a, The effect of display size and paragraph splitting on reading lengthy text from screen, *Behaviour and Information Technology*, 9(3), 215–27.

Dillon, A., Richardson, J. and McKnight, C., 1990b, Navigation in hypertext: a critical review of the concept, in Diaper, D., Gilmore, D., Cockton, G. and Shackel, B. (Eds) *INTERACT '90*, North Holland: Amsterdam, 587–92.

Dillon, A., Richardson, J. and McKnight, C., 1991, Institutionalising human factors in the design process: the ADONIS experience, *Contemporary Ergonomics '91*, London: Taylor and Francis, 421–6.

Dillon, A., Richardson, J. and McKnight, C., 1993b, Space – the final chapter: or why physical representations are not semantic intentions. In C. McKnight, A. Dillon and J. Richardson (Eds.) *Hypertext: A Psychological Perspective*. Chicester: Ellis Horwood, 169–192.

Dillon, A. and Sweeney, M., 1988, The application of cognitive psychology to CAD. In D, Jones and R. Winder (Eds.), *People and Computers IV*. Cambridge: Cambridge University Press, 477–488.

Dillon, A., Sweeney, M. and Maguire, M., 1993, A survey of usability evaluation practices and requirements in the European IT industry, in Alty, J., Guest, S. and Diaper, D. (Eds) *HCI '93. People and Computers VII*, Cambridge: Cambridge University Press.

Dillon, A., Sweeney, M., Herring, V., John, P. and Fallon, E., 1988, The psychology of designer style, *The Alvey Conference 1988*, DTI/IED Publications, 271–5.

Downs, R. and Stea, D. (Eds) 1974, *Image and Environment: Cognitive Mapping and Spatial Behaviour*, London: Edward Arnold.

Duchnicky, R. L. and Kolers, P. A., 1983, Readability of text scrolled on a visual display terminal as a function of window size, *Human Factors*, **25**(6), 683–92.

Duffy, T., Palmer, J. and Mehlenacher, B., 1992, *On-Line Help: Design and Evaluation*, Norwood NJ: Ablex.

Dunn, R., 1984, *Software Defect Removal*, New York: McGraw-Hill.

Eason, K., 1988, *Information Technology and Organisational Change*, London: Taylor and Francis.

Eason, K., Harker, S. and Poulson, D., 1986, Preliminary investigations into the use of human factors data in the design process, *HUSAT Memo No. 377*, Loughborough University of Technology (available on request).

Easteal, C. and Davies, G., 1989, *Software Engineering: Analysis and Design*, London: McGraw-Hill.

Edwards, D. and Hardman, L., 1989, 'Lost in Hyperspace': cognitive mapping and navigation in a hypertext environment, in McAleese, R. (Ed.) *Hypertext: Theory into Practice*, Oxford: Intellect, 105-25.

Egan, D., Remde, J., Landauer, T., Lochbaum, C. and Gomez, L., 1989, Behavioural evalution and analysis of a hypertext browser, *Proceedings of CHI '89*, New York: Association of Computing Machinery, 205-10.

Elkerton, J. and Williges, R., 1984, Information retrieval strategies in a file search environment, *Human Factors*, 26(2), 171-84.

Ellis, A., 1984, *Reading, Writing & Dyslexia: a cognitive analysis*, Hillsdale NJ: LEA.

Elm, W. and Woods, D., 1985, Getting lost: a case study in interface design, *Proceedings of the Human Factors Society 29th Annual Meeting*, Santa Monica CA: Human Factors Society, 927-31.

Engelbart, D., 1963, A conceptual framework for the augmentation of man's intellect, in Howerton, P. and Weeks, D. (Eds) *Vistas in Information Handling*, Vol. 1, London: Spartan Books, 1-29.

Ericsson, K. A., and Simon, H. A., 1984, *Protocol Analysis*, Cambridge, MA: MIT Press.

Ewing, J., Mehrabanzad, S., Sheck, S., Ostroff, D. and Shneiderman, B., 1986, An experimental comparison of a mouse and arrow-jump keys for an interactive encyclopedia, *International Journal of Man–Machine Studies*, 24(1), 29-45.

Feldman, T., 1990, *The Emergence of the Electronic Book*, British National Bibliography Research Fund Report 46, The British Library.

Ferguson, G., 1959, *Statistical Analysis in Psychology and Education*, New York: McGraw-Hill.

Fleishman, E. and Quaintance, M., 1984, *Taxonomies of Human Performance*, Orlando FL: Academic Press.

Forester, T. (Ed.) 1985, *The Information Technology Revolution*, Oxford: Blackwell.

Galer, M. and Taylor, B., 1989, Human Factors in Information Technology: Esprit Project 385, in Megaw, E. (Ed.) *Contemporary Ergonomics 1989*, London: Taylor and Francis, 82-6.

Gardiner, M. and Christie, B. (Eds) 1987, *Applying Cognitive Psychology to User-Interface Design*, Chichester: John Wiley and Sons.

Gardner, A. and McKenzie, J., 1988, *Human Factors Guidelines for the Design of Computer-Based Systems*, London: DTI/MOD. Available from HUSAT Research Institute, Loughborough University of Technology.

Garland, J., 1982, *Ken Garland and Associates: Designers – 20 years work and play*. Cited in, Waller, R. (1987) The typographic contribution to language: towards a model of typographic genres and their underlying structures. PhD Thesis, Dept. of Typography and Graphic Communication, University of Reading.

Garnham, A., 1987, *Mental Models as Representations of Text and Discourse*, Chichester: Ellis Horwood.

Girill, T. and Luk, C., 1983, Document: An interactive online solution to four documentation problems, *Communications of the ACM*, 26(5), 328-37.

Girill, T., Luk, C. and Norton, S., 1987, Reading patterns in online documentations: How transcript analysis reflects text design, software constraints and user preferences, in *Proc. of 34th International Technical Communications Conference*, Washington, DC: STC. 111-14.

Gordon, S., Gustavel, J., Moore, J. and Hankey, J., 1988, The effects of hypertext on reader knowledge representation, *Proceedings of the Human Factors Society 32nd Annual Meeting*, Santa Monica CA: Human Factors Society, 296–300.

Gould, J. D. and Grischkowsky, N., 1984, Doing the same work with hard copy and cathode-ray tube (CRT) computer terminals, *Human Factors*, **26**(3), 323–37.

Gould, J. and Grischkowsky, N., 1986, Does visual angle of a line of characters affect reading speed? *Human Factors*, **28**(2), 165–173.

Gould, J. D., Alfaro, L., Barnes, V., Finn, R., Grischkowsky, N. and Minuto, A., 1987a, Reading is slower from CRT displays than from paper: attempts to isolate a single variable explanation, *Human Factors*, **29**(3), 269–99.

Gould, J. D., Alfaro, L., Finn, R., Haupt, B. and Minuto, A., 1987b, Reading from CRT displays can be as fast as reading from paper, *Human Factors*, **29**(5), 497–517.

Hammond, N. and Allinson, L., 1987, The travel metaphor as design principle and training aid for navigating around complex systems, in Diaper, D. and Winder, R. (Eds) *People and Computers III*, Cambridge: Cambridge University Press, 75–90.

Hammond, N. and Allison, L., 1989, Extending hypertext for learning: an investigation of access and guidance tools, in Sutcliffe, A. and Macaulay, L. (Eds) *People and Computers V*, Cambridge: Cambridge University Press, 293–304.

Hammond, N., Jorgensen, A., Maclean, A., Barnard, P. and Long, J., 1983, Design practice and interface usability: evidence from interviews with designers, *IBM Hursley Human Factors Report HF 082*, Hursley Park, Winchester.

Hannigan, S. and Herring, V., 1986, The role of human factors in the design of IT products, Deliverable A12b, ESPRIT Project 385 – HUFIT, HUSAT Research Institute, Loughborough.

Harrison, M. and Thimbleby, H. (Eds) 1990, *Formal Methods in Human Computer Interaction*, Cambridge: Cambridge University Press.

Hartley, J., 1985, *Designing Instructional Text*, 2nd Edn, London: Kogan Page.

Hassard, J., 1988, FOCUS as a phenomenological technique for job analysis: its use in multiple paradigm research, *International Journal of Man–Machine Studies*, **27**, 413–33.

Hatt, F., 1976, *The Reading Process*, London: Clive Bingley.

Hayes-Roth, B., 1983, The blackboard architecture: a general framework for problem-solving? *HPP Report No. HPP-83-30*, Stanford University, Dept. of Computer Science.

Helander, M. G., Billingsley, P. A. and Schurick, J. M., 1984, An evaluation of human factors research on visual display terminals in the workplace, in Muckler, F. (Ed.) *Human Factors Review: 1984*, Santa Monica CA: Human Factors Society, 55–129.

Horney, M., 1993, A measure of hypertext linearity, *Journal of Educational Multimedia and Hypermedia*, **2**(1), 67–82.

Hudson, L., 1968, *Frames of Mind*, Penguin: Harmondsworth.

Huey, E. B., 1908, *The Psychology and Pedagogy of Reading*, New York: Macmillan.

Jaschinski-Kruza, W., 1990, On the preferred viewing distances to screen and document at VDU workplaces, *Ergonomics*, **33**(8), 1055–63.

Johnson-Laird, P., 1983, *Mental Models*, Cambridge: Cambridge University Press.

Jonassen, D., 1982, *The Technology of Text, Vol I. Principles for Structuring, Designing, and Displaying Text*, Englewood Cliffs NJ: Educational Technology Publications.

Jones, W. P. and Dumais, S. T., 1986, The spatial metaphor for user interfaces: experimental tests of reference by location versus name, *ACM Transactions on Office Information Systems*, **4**(1), 42–63.

Just, M. A. and Carpenter, P., 1980, A theory of reading: from eye movements to comprehension, *Psychological Review*, **87**(4), 329–54.

Kak, A. V., 1981, Relationships between readability of printed and CRT-displayed text, *Proceedings of Human Factors Society – 25th Annual Meeting*, Santa Monica CA: Human Factors Society, 137–40.

Karat, C., Campbell, R. and Fiegel, T., 1992, Comparison of empirical testing and walkthrough methods in user interface design, *Proceedings of CHI '92*, New York: ACM Press, 397–404.

Kelly, G., 1955, *The Psychology of Personal Constructs*, 2 Vols., New York: Norton.

Kerlinger, F., 1973, *Foundations of Behavioral Research*, New York: Holt, Rinehart and Winston.

Kerr, S. T., 1986, Learning to use electronic text: an agenda for research on typography, graphics, and interpanel navigation, *Information Design Journal*, 4(3), 206–11.

Kieras, D. and Polson, P., 1985, An approach to the formal analysis of user complexity, *International Journal of Man-Machine Studies*, 22, 365–94.

Kintsch, W., 1974, *The Representation of Meaning in Memory*, Hillsdale NJ: Lawrence Erlbaum Associates.

Kintsch, W. and Yarborough, J., 1982, The role of rhetorical structure in text comprehension, *Journal of Educational Psychology*, 74, 828–34.

Kline, P., 1988, *Psychology Exposed, or The Emperor's New Clothes*, London: Routledge.

Kolers, P. A., Duchnicky, R. L., and Ferguson, D. C., 1981, Eye movement measurement of readability of CRT displays, *Human Factors*, 23(5), 517–27.

Kruk, R. S. and Muter, P., 1984, Reading continuous text on video screens, *Human Factors*, 26(3), 339–45.

Kuhn, T. S., 1962, *The Structure of Scientific Revolutions*, Chicago: University of Chicago Press.

Landauer, T., 1987, Relations between cognitive psychology and computer systems design, in Carroll, J. (Ed.) *Interfacing Thought*, Cambridge MA: MIT Press, 1–25.

Lawson, B., 1979, *How Designers Think*, London: Architectural Press.

Leventhal, L., Teasley, B., Instone, K., Rohlman, D. and Farhat, J., 1993, Sleuthing in HyperHolmes: an evaluation of using hypertext versus a book to answer questions, *Behaviour and Information Technology*, 12(3), 149–64.

Licklider, J., 1965, *Libraries of the Future*, Cambridge MA: MIT Press.

Lovelace, E. A. and Southall, S. D., 1983, Memory for words in prose and their locations on the page, *Memory and Cognition*, 11(5), 429–34.

Mack, R. L., Lewis, C. H. and Carroll, J. M., 1983, Learning to use word-processors: problems and prospects, in *ACM Transactions on Office Information Systems*, 1(3), 254–71.

Mandler, J. and Johnson, N., 1977, Remembrance of things parsed: story structure and recall, *Cognitive Psychology*, 9, 111–51.

Martin, A., 1972, A new keyboard layout, *Applied Ergonomics*, 3(1), 42–51.

McAleese, R., 1989, Navigation and browsing in hypertext, in McAleese, R. (Ed.) *Hypertext: Theory into Practice*, Oxford: Intellect, 6–44.

McAleese, R. and Green, C., 1990, *Hypertext: State of the Art*, Oxford: Intellect.

McClelland, J. L., and Rumelhart, D., 1981, An interactive activation model of context effects in letter perception: Part 1. an account of basic findings, *Psychological Review*, 88, 375–407.

McConkie, G., Underwood, N., Zola, D. and Wolverton, G., 1985, Some temporal characteristics of processing during reading, *Journal of Experimental Psychology: Human Perceptions and Performance*, 11(2), 168–86.

McKnight, C., 1981, Subjectivity in sentencing, *Law and Human Behavior*, 5(2/3), 141–7.

McKnight, C., Dillon, A. and Richardson, J., 1990a, A comparison of linear and hypertext formats in information retrieval, in McAleese, R. and Green, C. (Eds) *Hypertext: State of the Art*, Oxford: Intellect, 10–19.

McKnight, C., Dillon, A. and Richardson, J., 1990b, 'Project CHIRO', HUSAT Research Institute, Loughborough University of Technology, BLR&DD Research Report.

McKnight, C., Dillon, A. and Richardson, J., 1991, *Hypertext in Context*, Cambridge: Cambridge University Press.

McKnight, C., Richardson, J. and Dillon, A., 1989, The authoring of hypertext documents, in McAleese, R. (Ed.) *Hypertext: Theory into Practice*, Oxford: Intellect, 138–47.

McKnight, C., Dillon, A., Richardson, J., Haraldsson, H. and Spinks, R., 1992, Information access in different media: an experimental comparison. In E. J. Lovesey (Ed.) *Contemporary Ergonomics 1992*, London: Taylor and Francis, 515–519.

Meadows, A. J., 1974, *Communication in Science*, London: Butterworth.

Medawar, P., 1964, Is the scientific paper a fraud? in Edge, D. (Ed.) *Experiment*, London: BBC Publications.

Miller, R. B., 1967, Task taxonomy: science or technology? in Singleton, W., Easterby, R. and Whitfield, D. (Eds) *The Human Operator in Complex Systems*, London: Taylor and Francis.

Mills, C. B. and Weldon, L. J., 1986, Reading text from computer screens, *ACM Computing Surveys*, **19**(4).

Milner, N., 1988, A review of human performance and preference with different input devices to computer systems, in Jones, D. and Winder, R. (Eds) *People and Computers IV*, Cambridge: Cambridge University Press, 341–62.

Mitchell, D., 1982, The Process of Reading: a cognitive analysis of fluent reading and learning to read. New York: Wiley.

Monk, A., Walsh, P. and Dix, A., 1988, A comparison of hypertext, scrolling, and folding as mechanisms for program browsing, in Jones, D. and Winder, R. (Eds) *People and Computers IV*, Cambridge: Cambridge University Press, 421–35.

Moran, T., 1981, The command language grammar, a representation for the user interface of interactive computer systems, *International Journal of Man–Machine Studies*, **15**(1), 3–50.

Muter, P. and Maurutto, P., 1991, Reading and skimming from computer screens and books: the paperless office revisited? *Behaviour and Information Technology*, **10**(4), 257–266.

Muter, P., Latremouille, S. A., Treurniet, W. C., and Beam, P., 1982, Extended reading of continuous text on television screens, *Human Factors*, **24**(5), 501–8.

Neal, A. and Darnell, M., 1984, Text editing performance with partial line, partial page and full page displays, *Human Factors*, **26**(4), 431–41.

Nelson, T., 1987, *Literary Machines*, Abridged Electronic Version 87.1 San Antonio: Ted Nelson.

Newell, A. and Card, S., 1985, The prospects for psychological science in Human–Computer Interaction, *Human–Computer Interaction*, **1**, 209–42.

Nielsen, J., 1990, Hypertext/Hypermedia. New York: Academic Press.

Nielsen, J., 1992, Finding usability problems through heuristic evaluation, in *Proceedings of CHI '92*, New York: ACM, 373–80.

Niijar, H., 1993, 'Reader's perceptions of structure in text', unpublished MSc. thesis, Dept. of Computer Studies, Loughborough University of Technology.

Nisbett, R. and Wilson, T., 1977, Telling more than we can know: verbal reports on mental processes, *Psychological Review*, **84**, 231–59.

Norman, D., 1986, Cognitive Engineering, in Norman, D., and Draper, S. (Eds) *User Centred System Design*, Hillsdale NJ: Lawrence Erlbaum Associates, 31–61.

Norman, D., 1988, The Psychology of Everyday Things. New York: Basic Books.

Norman, D. and Draper, S. (Eds), *User Centred System Design*, Hillsdale NJ: Lawrence Erlbaum Associates.

Norman, K. and Chen, J., 1988, The effect of tree structure on search in a hierarchical menu selectration system, *Behaviour and Information Technology*, **7**(1), 51–65.

Oborne, D. and Holton, D., 1988, Reading from screen versus paper: there is no difference, *International Journal of Man–Machine Studies*, **28**(1), 1–9.

Olshavsky, J., 1977, Reading as problem solving: an investigation of strategies, *Reading Research Quarterly*, **4**, 654–74.

Oppenheim, A. N., 1966, *Questionnaire Design and Attitude Measurement*, London: Heinemann.

Pask, G., 1976, Styles and strategies of learning, *British Journal of Education Psychology*, **46**, 128–48.

Pastoor, S., Schwartz, E., and Beldie, I. P., 1983, The relative suitability of four dot-matrix sizes for text presentation on colour television screens, *Human Factors*, **25**(3), 265–72.

Pearce, B. (Ed.) 1984, *Health Hazards of VDUs?*, Chichester: John Wiley and Sons.

Pearson, R. G. and Byars, G. E., 1956, 'The development and validation of a checklist for measuring subjective fatigue', Report #TR-56-115, San Antonio, TX: USAF School of Aviation Medicine.

Polson, P., Muncher, E. and Engelbeck, G., 1986, A test of a common elements theory of transfer, in Mantei, M. and Orbeton, P. (Eds) *Proceedings of CHI '86*, New York: Association for Computing Machinery, 78–83.

Polson, P., Muncher, E. and Kieras, D., 1987, 'Transfer of skills between inconsistent editors'. Technical Report No. 87-10, Boulder: University of Colorado, Institute of Cognitive Science.

Popper, K., 1972, *The Logic of Scientific Discovery*, 3rd Edn., London: Hutchinson.

Popper, K., 1986, *The Poverty of Historicism*, Reading: ARK (originally published in 1957).

Pugh, A., 1975, The development of silent reading, in Latham, W. (Ed.) *The Road to Effective Reading*, London: Ward Lock.

Pugh, A., 1979, Styles and strategies in adult silent reading, in Kolers, P., Wrolstad, M. and Bouma, H. (Eds) *Processing of Visible Language 1*, London: Plenum Press.

Pullinger, D., 1984, Design and presentation of the CHF journal on the BLEND system, *Visible Language*, **18**(2), 171–85.

Radl, G. W., 1980, Experimental investigations for optimal presentation mode and colours of symbols on the CRT screen, in Grandjean, E. and Vigliani, E. (Eds) *Ergonomic Aspects of Visual Display Terminals*, London: Taylor and Francis, 127–37.

Rasmussen, J., 1986, *Information Processing and Human–Machine Interaction: An Approach to Cognitive Engineering*, London: North Holland.

Richardson, J., Dillon, A. and McKnight, C., 1989, The effect of window size on reading and manipulating electronic text, in Megaw, E. (Ed.) *Contemporary Ergonomics 1989*, London: Taylor and Francis, 474–9.

Richardson, J., Dillon, A., McKnight, C. and Saadat-Samardi, M., 1988, 'The manipulation of screen presented text: experimental investigation of an interface incorporating a movement grammar', HUSAT memo #431, Loughborough University of Technology.

Rivlin, C., Lewis, R. and Cooper, R., 1990, *Guidelines for Screen Design*, Oxford: Blackwell Scientific.

Rothkopf, E. Z., 1971, Incidental memory for location of information in text, *Journal of Verbal Learning and Verbal Behavior*, **10**, 608–13.

Rumelhart, D., 1977, Toward an interactive model of reading, in Dornic, S. (Ed.) *Attention and Performance VI*, Hillsdale NJ: Erlbaum.

Ryle, A., 1976, Some clinical applications of grid technique, in Slater, P. (Ed.) *The Measurement of Intrapersonal Space by Grid Technique*, 2 Vols., London: John Wiley and Sons.

Samuels, S. and Kamil, M., 1984, Models of the reading process, in Pearson, P. (Ed.) *Handbook of Reading Research*, New York: Longman, 185–224.

Sauter, S., Gottlieb, M., Rohrer, K. and Dodson, V., 1983, 'The well-being of video display terminal users: an exploratory study', Report No: 210-79-0034, Cincinnati, OH: US Dept. of Health and Human Sciences.

Schumacher, G. and Waller, R., 1985, Testing design alternatives: a comparison of procedures, in Duffy, T. and Waller, R. (Eds) *Designing Usable Texts*, Orlando FL: Academic Press, 377–403.

Schwartz, E., Beldie, I. and Pastoor, S., 1983, A comparison of paging and scrolling for changing screen contents by inexperienced users, *Human Factors*, 25, 279–82.

Shackel, B., 1959, Ergonomics for a computer, *Design*, 120, 36–9.

Shackel, B., 1986, Ergonomics in design and usability, in Harrison, M. and Monk, A. (Eds) *People and Computers: Designing for Usability*, Cambridge: Cambridge University Press.

Shackel, B., 1987, An overview of research on electronic journals, in Salvendy, G. (Ed.) *Cognitive Engineering in the Design of Human Computer Interaction and Expert Systems*, Amsterdam: Elsevier, 193–206.

Shackel, B., 1990, BLEND-9: Final Report and Overview. British Library Research and Development Dept., London.

Shackel, B., 1991, Usability – context, framework, definition, design and evaluation, in Shackel, B. and Richardson, S. (Eds) *Human Factors for Informatics Usability*, Cambridge: Cambridge University Press, 21–37.

Sharratt, B., 1987, The incorporation of early interface evaluation into command language grammar specifications, in Diaper, D. and Winder, R. (Eds) *People and Computers III*, Cambridge: Cambridge University Press.

Shaw, M. L. G., 1980, *On Becoming a Personal Scientist*, London: Academic Press.

Shaw, M. L. G. and Gaines, B., 1987, KITTEN: Knowledge initiation and transfer tools for experts and novices, *International Journal of Man–Machine Studies*, 27, 251–80.

Sheridan, T., Senders, J., Moray, N., Stoklosa, J., Guillame, J. and Makepeace, D., 1981, 'Experimentation with a Multidisciplinary Teleconference and Electronic Journal on Mental Workload', Unpublished report to the National Science Foundation (Division of Science Information Access Improvements), June.

Shneiderman, B., 1984, The future of interactive systems and the emergence of direct manipulation, in Vassiliou, Y. (Ed.) *Human Factors and Interactive Computer Systems*, Norwood NJ: Ablex, 1–27.

Shneiderman, B., 1987, *Designing the User Interface: Strategies for Effective Human-Computer Interaction*, San Francisco: Addison Wesley.

Simpson, A., 1989, Navigation in hypertext: design issues, *International OnLine Conference '89*, London, December.

Simpson, A., 1990, 'Towards the design of an electronic journal', unpublished PhD thesis, Dept. of Human Sciences, Loughborough University of Technology.

Slater, P., 1976, *The Measurement of Intrapersonal Space by Grid Technique*, 2 Vols, London: John Wiley and Sons.

Smedshammar, H., Frenckner, K., Nordquist, C. and Romberger, S., 1989, Why is the difference in reading speed when reading from VDUs and from paper bigger for fast readers than for slow readers? Paper presented at *WWDU 1989. Second International Scientific Conference*, Montreal.

Smith, A. and Savory, M., 1989, Effects and after-effects of working at a VDU: investigation of the influence of personal variables. In: E. D. Megaw (ed.) *Contemporary Ergonomics 1989*. London: Taylor and Francis.

Smith, F., 1978, *Reading*, Cambridge: Cambridge University Press.

Spinks, R., 1991, 'An experimental comparison of hypertext and paper for information location in lengthy text', Unpublished MSc. thesis, Dept. of Human Sciences, Loughborough University of Technology.

Spring, M., 1991, *Electronic Printing and Publishing: The Document Processing Revolution*, New York: Marcel Dekker, Inc.

Stark, H., 1990, A comparison of jump and pop-up windows in hypertext. In: McAleese, R. and Green, C. (Eds) *Hypertext: State of the Art*, Oxford: Intellect.

Starr, S. J., 1984, Effects of video display terminals in a business office, *Human Factors*, 26(3), 347–56.

Stevens, G., 1983, User-friendly computer systems?: a critical examination of the concept, *Behaviour and Information Technology*, 2(1), 3–16.

Suchman, L., 1988, *Plans and Situated Action*, Cambridge: Cambridge University Press.

Sweeney, M., Maguire, M. and Shackel, B., 1993, Evaluating user–computer interaction: a framework, *International Journal of Man–Machine Studies*, 38(4), 689–712.

Tinker, M. A., 1958, Recent studies of eye movements in reading, *Psychological Bulletin*, 55, 215–31.

Tinker, M. A., 1963, *Legibility of Print*, Ames, Iowa: Iowa State University Press.

Thorndyke, P. and Hayes-Roth, B., 1982, Differences in spatial knowledge acquired from maps and navigation. *Cognitive Psychology*, 14, 560–589.

Tombaugh, J., Lickorish, A. and Wright, P., 1987, Multi-window displays for readers of lengthy texts, *International Journal of Man–Machine Studies*, 26, 597–616.

Tovey, M., 1986, Designing with both halves of the brain, *Design Studies*, 5(4), 219–28.

Trigg, R. and Suchman, L., 1989, Collaborative writing in NoteCards, in McAleese, R. (Ed.) *Hypertext: Theory into Practice*, Norwood NJ: Ablex, 45–61.

Tuck, B., McKnight, C., Hayet, M. and Archer, D., 1990, *Project Quartet*, Library and Information Research Report 76, London: The British Library.

van Dijk, T. A., 1980, *Macrostructures*, Hillsdale NJ: Lawrence Erlbaum Associates.

van Dijk, T. A. and Kintsch, W., 1983, *Strategies of Discourse Comprehension*, London: Academic Press.

Venezky, R. L., 1984, The history of reading research, in Pearson, P. (Ed.) *Handbook of Reading Research*, New York: Longman, 3–38.

Ventura, C., 1988, Why switch from paper to electronic manuals? *Proceedings of the ACM Conference on Document Processing Systems*, Santa Fe NM: Association for Computing Machinery, 111–16.

Waern, Y. and Rollenhagen, C., 1983, Reading text from visual display units (VDUs), *International Journal of Man–Machine Studies*, 18, 441–65.

Waller, R., 1984, Designing government forms: a case study, *Information Design Journal*, 4, 36–57.

Waller, R., 1986, What electronic books will have to be better than, *Information Design Journal*, 5, 72–5.

Waller, R., 1987, 'The typographic contribution to language: towards a model of typographic genres and their underlying structures'. Unpublished PhD thesis, Dept. of Typography and Graphic Communication, University of Reading.

Weinberg, G., 1971, *The Psychology of Computer Programming*, New York: Van Nostrand Reinhold.

Whalley, P. and Fleming, R., 1975, An experiment with a simple recorder of reading behaviour, *Programmed Learning and Educational Technology*, 12, 120–4.

Wharton, C., Bradford, J., Jeffries, R. and Franzke, M., 1992, Applying Cognitive Walkthroughs to more complex user interfaces: experiences, issues and recommendations. *Proceedings of CHI '92*, New York: ACM, 381–8.

Wetherell, A., 1979, Short-term memory for verbal and graphic route information. *Proceedings of the Human Factors Society 23rd Annual Meeting*, Santa Monica CA: Human Factors Society.

Whitefield, A., 1989, Constructing appropriate models of computer users: the case of engineering designers, in Long, J. and Whitefield, A. (Eds) *Cognitive Ergonomics and Human Computer Interaction*, Cambridge: Cambridge University Press, 66–94.

Wickens, C., 1984, *Engineering Psychology and Human Performance*, Columbus: Charles Merrill.

Wilkinson, R. T. and Robinshaw, H. M., 1987, Proof-reading: VDU and paper text compared for speed, accuracy and fatigue, *Behaviour and Information Technology*, 6(2), 125–33.

Wilson, M.D., Barnard, P.J. and Maclean, A., 1986, 'Task analysis in human–computer interaction'. Hursley Human Factors Laboratory Report HF122, Hursley Park, Winchester.

Winograd, T. and Flores, F., 1988, *Understanding Computers and Cognition*, Reading MA: Addison Wesley.

Witkin, H.A., Moore, C.A., Goodenough, D.R. and Cox, P.W., 1977, Field dependent and field independent cognitive styles and their educational implications, *Review of Educational Research*, 1–64.

Wittgenstein, L., 1953, *Philosophical Investigations*, New York: MacMillan.

Wright, P., 1980, Textual literacy: an outline sketch of psychological research on reading and writing, in Kolers, P., Wrolstad, M. and Bouma, H. (Eds) *Processing of Visible Language 2*, London: Plenum Press.

Wright, P. and Lickorish, A., 1983, Proof-reading texts on screen and paper, *Behaviour and Information Technology*, 2(3), 227–35.

Wright, P. and Lickorish, A., 1988, Colour cues as location aids in lengthy texts on screen and paper, *Behaviour and Information Technology*, 7(1), 11–30.

Zechmeister, E. and McKillip, J., 1972, Recall of place on a page, *Journal of Educational Psychology*, **63**, 446–53.

Zechmeister, E., McKillip, J., Pasco, S. and Bespalec, D., 1975, Visual memory for place on the page, *Journal of General Psychology*, **92**, 43–52.

Appendix

Example protocol for reader in validity experiment (Word Processor User)

Time	Comment	Action
0.00		Reads question 1
0.11	I'm going to the Index to see if there's anything on taste	Scrolls
0.17	No...Contents	Reads Contents
0.24	No...I've a feeling Introduction covers the taste of wine...I'll check that	Scrolls down to Introduction
0.31		Scrolls further and reads
0.45	...about the colour?	
0.48		Scrolls further and reads
0.53		Scrolls further and reads
0.58		Scrolls further and reads
1.03		Scrolls further and reads
1.06		Has reached new section
1.08		Still reading
1.17	Right I think sweetness is one...	Writes down 'Sweetness'
1.20		Scrolls on and reads
1.31		Reaches new section
1.41	I think it's Sweetness and Body, just these two...	Scanning text, writes down 'Body'
1.52	I'll just check to see if there's anything later on...	Scrolling and reading further
1.55		Reaches new section
1.58		Reaches new section
2.03	No...I don't think so...I'm going to leave that question...	
2.08		Reads question 2
2.18	I've got a feeling I've just seen that when I was looking for...fermentation	Scrolls directly back to top of document and reads the Contents

2.23	Basically the yeast dies . . .	Scroll to relevant section and scans text
2.37		Confirms answer and writes it down
3.04		Reads question 3
3.10	Something to do with Fermentation? . . .	Scans text, then scrolls down
3.19	Yes . . . to keep the yeast alive and stop the wine burning . . .	Reads text
3.21		Writes down answer
3.41		Reads question 4
3.47	Again, I think I've seen something on that . . .	Scrolls continually down while scanning very quickly
3.55		Scrolls further down
4.00		Still scrolling and scanning rapidly
4.07		Still scrolling, has moved into previously unread text
4.11	Ah . . . I must have passed it . . .	
4.19		Scrolls back through the text scanning rapidly
4.28		Has scrolled back as far as Fermentation section
4.32		Has scrolled back to Introduction
4.37		Reading Contents
4.44	There's nothing in the Contents that's telling me . . . but I'm sure it must be near the start of the document . . . It's fundamental . . .	Reading the early part of the Introduction
4.53	Something to do with colour maybe . . .	Still reading Introduction and following sections Scrolling down as necessary
5.10	But I don't know what you mean by 'lighter' (a ref to the question) or 'taste'	Scrolls further down the introduction
5.23	. . . Ah . . . it's the caramelization of residual sugar	Quotes from the text having found a suitable answer
5.32		Writes down answer
5.51		Reads question 5
6.07	To the Index then . . . I haven't seen anything on this	Scrolls directly down to the bottom of the document
6.11		Scrolls slowly back up to the top of the Index

6.17		Scrolls quickly to the body of the Index
6.19	Grapes on page 1? . . . No . . .	Reading Index terms
6.25		Scrolls directly up to the top of the file and then scrolls slower down to a section in the introduction
6.56		Starts scrolling back through the Introduction
7.01	It must be in the body of the report then . . .	Reading section on Fermentation again
7.03		Scrolls down to Aging section
7.06		Scrolling and reading the following sections
7.21		Studying the text intensely
7.27		Reading sections on Sweetness and Body Scrolling slowly as required
7.31		Reading section on wine categories: table and dessert wines
7.33	Oh . . . dessert wines	Writes down answer
7.47		Reads question 6
7.51	I've just passed a section on aging	Scrolls up to Aging
7.55		Reads through section
8.09	Mentions a bit about vintage port . . . doesn't say how old it should be though . . .	
8.22	I think I'll find the section on Port	Goes straight up to Contents
8.25		Browsing through Contents
8.29	No . . . Index	Drags scroll bar down to end
8.38		Drags scroll bar to top
8.43		Selects Goto command from the menu. Inputs Goto Page 4
8.45		Views sections on Table and Dessert wines
8.51		Scrolls down to section on Aperitifs and Fortified wines
8.53	Port . . .	Finds relevant reference
8.59	Vintage port . . . at least 20 years old	Writes down answer
9.14		Reads question 7
9.17	Haven't seen anything on this method before . . . Solera . . . Check the index	Drags scroll bar down to end

9.23	What a useless index	
9.36		Drags scroll bar back up to middle of text
9.42		Reads serially through the text from section on Aging to section on sparkling wines, using slow scroll as necessary
10.30		Drags scroll bar to top to see Contents
10.34	I've just remembered . . . I can search for . . .	Invokes search facilities
10.36		Inputs 'Solera'
10.52		Finds the appropriate answer
10.54	Oh . . . I missed that . . . I skimmed past it	Writes down answer
11.03		Reads question 8
11.07		Drags scroll bar to top to read Contents
11.09	My god . . . I'll search for that again.	Invokes search facilities
11.11		Inputs 'Woodworm'
11.21	'Continue from beginning . . . ?' yes	Hits return
11.25		End of document message Search is unsuccessful
11.27		Reads question again
11.35	Wormwood . . . bloody hell!	Corrects search term
11.55		Starts search
12.01	Vermouth eh . . .	Term is found in relevant section
12.03		Writes down answer
12.10		Reads question 9
12.12	That's got something to do with champagne wines . . .	There's a reference to sparkling wines at present position in text. Reads this
12.24		Scrolls down text Continues reading
12.29	Produces natural effervescence . . .	Writes down answer
12.51		Reads question 10
12.54		Scrolls directly to top of text to see Contents
12.57		Scrolls slowly through contents while reading it
13.03		Invokes Goto facility Inputs '10'
13.07		Reading section on Rhone
13.20	Mainly red wines are there . . .	Reads through to Loire section

13.25	Oh yeah . . . mainly white . . . a bit of luck	Writes down answer
13.38		Reads question 11
13.44		Scrolls directly to top to view Contents
13.50	Countries start on page 6 . . .	
13.54		Invokes Goto command Inputs '6'
14.01		Reading section on France
14.04	France, fairly obviously . . .	Reads through relevant sections on countries
14.24	France and the US, that is in the Italy section . . .	Writes down answer
14.39		Reads question 12
14.42	Types of wine? . . .	Scrolls slowly back through the text
15.00		Scrolls slowly back reading text as he does so
15.47		Reaches Contents
15.52	Ah . . . aging on page 2?	Scrolls back down
15.59		Reads section on aging
16.18	OK . . . it must be those . . .	Makes notes
16.55	I think that must be the reference to it but it's not the same as 'as soon as it's bottled' . . .	Refers to question
17.10	Champagne . . . if you naturally ferment it, it's just going to stay in the bottle . . .	
17.20		Jumps down to Index
17.24		Scrolls slowly back through the Index
17.30	What about Beaujolais? . . .	
17.42		Invokes Goto facility and inputs '11'
17.53		Read sections on Burgundy and Beaujolais
18.20		Scrolls back on section just read
18.24	I'll look for bottled or something . . .	Invokes Search facilities and inputs 'bottle'
18.31	No . . .	Cancels this action
18.38		Reads sections on Loire and Burgundy
18.43		Invokes search facilities again and searches on 'bottle'

18.48		Prompted to continue search from beginning of file he cancels
18.52	No ...	Tries search again using same term
19.06	No ... this isn't it ...	Has found numerous references to 'bottle' in section on Aging
19.13		Cancels Find command
19.19		Invokes search facilities again and searches on same term
19.21		Reads the section on Port and Sherry where search facilities have taken him
19.35		Find Next
19.37	Germany and Italy?	Find Next
19.41		In Champagne section. Find Next
19.45		Is in California section. Find Next then takes him to Sparkling Wines section
19.50		Find Next returns the 'start from beginning message'. He cancels the Find command
19.54	So it's just champagne and semi-sweet, but that doesn't seem quite right ...	Writes down answer Session Ends.

Index

abstraction 65-9, 162-5
access to information 74, 88-9, 106, 165
  retrieval 28, 64
  validity experiment 144
accuracy of reading 63, 90, 116, 118,
  144
  screen and paper 32-4, 38, 58
ADONIS 63-5, 68-9, 115, 170
angle of viewing 46, 47, 159
anti-aliasing 49-50
aspect ratio 44

BLEND 88, 115

catalogues 5, 73-9, 81-2, 85
character size 31, 48-9
character spacing 48-9
classifying text 8, 72-9, 81-6, 103, 137,
  163-4
cloze test 108
Cognitive Complexity Theory (CCT) 66,
  132
comprehension 4, 52, 60-1, 90, 108-9,
  117, 165-6
  screen and paper 35-7, 38
computer aided design (CAD) 24
conference proceedings 77, 78-9, 82
constructs 76-82, 137-8
contents 4, 106, 112, 114
  manuals and journals 96, 97, 99, 100
  validity experiment 145-6, 148-9
costs 1, 11, 13, 16, 89, 121

dendograms 77, 80, 81
display characteristics 48-9, 50-1
  size 45, 52
document size 45, 52, 97, 101, 114,
  144-5

EIES project 88
ergonomics 12, 30, 74-5, 115, 166-7,
  171
  electronic text design 3-9, 121, 134,
  136

user-centred 18-19, 22, 23
product usability 14-16, 18-19
reading process 42, 51, 58, 65, 69
user-centred design 18-19, 22, 23
ESPRIT 171
eye movements 127-8, 132, 164
  reading process 38, 49-50, 52, 62, 90,
  105-6
eyestrain 34-5

fatigue 34-5, 38
feeling-tone checklist (FTC) 34
flicker 34-5, 46-7, 50-1, 159
FOCUS 77, 78-9
fonts 49-51, 159, 169
formatting 29
function keys 18, 40, 44

generation-conjecture-analysis model
  24-6
global schemata 110, 125-6
GOMS 67, 123, 132, 135, 168, 172
  task analysis 21
  usability 15
graphics 5, 15, 83, 95
GUIDE 53, 152, 154-5, 160

handbooks 88
human computer interaction (HCI)
  classifying text 74, 76, 84
  design 4, 6, 122-3, 131-3, 136, 168,
  172
  reading process 65, 66-8
  usability 12, 15, 136
  user-centred design 17, 18, 21, 23
  verbal protocols 92
human factors 12, 29, 162, 166, 171
  classifying text 74, 81, 82
  design 3-9, 103, 121-2, 131, 134, 136
  manuals 89
  reading process 62, 63-5, 68-70
  structure of text 109, 119
  task analysis 21, 22
  user-centred design 19, 22-3